Chemistry of Oil Recovery

Chemistry of Oil Recovery

**Robert T. Johansen and
Robert L. Berg,** EDITORS

United States Department of Energy

Based on a symposium

sponsored by the Division

of Petroleum Chemistry at

the 175th Meeting of the

American Chemical Society,

Anaheim, California, March

13–16, 1978.

ACS SYMPOSIUM SERIES **91**

AMERICAN CHEMICAL SOCIETY

WASHINGTON, D. C.   1979

Library of Congress CIP Data

Chemistry of oil recovery.
    (ACS symposium series; 91 ISSN 0097–6156)

    Includes bibliographies and index.

    1. Secondary oil recovery—Congresses.
    I. Johansen, Robert T. II. Berg, Robert L., 1949–   .
III. American Chemical Society. Division of Petroleum
Chemistry. IV. Series: American Chemical Society.
ACS symposium series; 91.

TN871.C464   622′.33′82   78-27298
ISBN 0-8412-0477-2   ACSMC8   91 1–182 1979

ACS Symposium Series

Robert F. Gould, *Editor*

FOREWORD

The ACS SYMPOSIUM SERIES was founded in 1974 to provide a medium for publishing symposia quickly in book form. The format of the Series parallels that of the continuing ADVANCES IN CHEMISTRY SERIES except that in order to save time the papers are not typeset but are reproduced as they are submitted by the authors in camera-ready form. Papers are reviewed under the supervision of the Editors with the assistance of the Series Advisory Board and are selected to maintain the integrity of the symposia; however, verbatim reproductions of previously published papers are not accepted. Both reviews and reports of research are acceptable since symposia may embrace both types of presentation.

CONTENTS

Contents

PREFACE

Understanding the chemistry of oil recovery is important for the development of new processes and for the improvement of old techniques for enhanced recovery. The chapters in this volume deal with the fundamental chemistry of the mechanisms involved in enhanced oil recovery.

Conventional (primary and secondary) recovery methods recover only a small fraction of the crude oil originally in place in a typical reservoir. The primary and secondary recovery techniques, which include pressure maintenance by gas injection and water flooding for improved recovery, leave approximately two-thirds of the original oil in the reservoir. As the conventional oil production of the United States continues to decline, enhanced oil recovery will play an important role in the utilization of our domestic resources. Conventional methods do not overcome the basic problems of oil being trapped within the rock pores and of the low mobility of the remaining oil.

Methods for increasing recovery of petroleum include thermal recovery processes, carbon dioxide injection, and chemical flooding. Thermal recovery methods use either in situ combustion or steam injection. In both cases, heat is added to the reservoir; consequently, the viscosity of the oil is reduced. The increased mobility of the oil within the reservoir allows significant additional production. The carbon dioxide injection process is a miscible displacement process in which carbon dioxide is injected into the reservoir where it forms a miscible mixture with the oil. The dissolved carbon dioxide volumetrically expands the oil and reduces its viscosity, thus allowing the oil to flow more readily thereby increasing its recovery. Chemical flooding includes techniques such as polymer flooding, caustic flooding, and micellar/polymer flooding. The polymer flood uses a mobility control agent, usually a polysaccharide or polyacrylamide, which is added to injection water to increase the viscosity of the displacement fluid. Sweep efficiency is improved and increased recovery results. The alkali added to the injection water in caustic flooding reacts in some manner with the oil and the formation and thereby increases recovery. Considerable disagreement over the actual mechanism(s) responsible for the improved recovery with caustic remains. Micellar/polymer flooding is the most recent development in enhanced oil recovery by chemical flooding. In this technique, a "slug" of a micellar fluid, consisting of surfactant, co-surfactant, oil, and brine, is injected into the formation. The interfacial tension is reduced, and oil is displaced towards the producing wells. A mobility control agent, usually a polymer, is injected behind the micellar material to prevent fingering and decomposition of the slug by the usually saline water which follows to drive the oil bank through the reservoir.

Considerable laboratory and field work is currently going on testing enhanced recovery techniques. In March, 1978, the *Oil and Gas Journal* (*1*) reported 196 active enhanced recovery projects with an approximate production of 370,000 barrels of oil per day. More than 150 additional projects are in the planning stages. In these projects alone, the vast potential of more than 450,000 barrels of oil per day from 150,000 acres exists. In order to obtain this production level, a large-scale expansion of production of chemicals for the enhanced recovery processes is required. For micellar/polymer flooding techniques, an increase of sulfonate production by a factor of four may be required (*2*). Polymer capacity may be required to expand by an even larger factor. Clearly a large increase in the production of chemicals will be required if these processes are to be implemented fully.

The chemistry of and physical mechanisms controlling the recovery of petroleum are better understood now than at any previous time; yet, an extremely large amount of work remains before recoveries can be predicted with certainty. Part of this uncertainty arises from the undetermined variables of the heterogeneous reservoir matrix from which oil is produced and partly from simply not understanding the total interrelationship of the forces acting on the fluids in the reservoir. New chemical techniques and ideas for measurement undoubtedly will develop from the kinds of studies reported in this publication: the result will be the achievement of a more efficient recovery of our petroleum resources.

Literature Cited

(1) Noran, Dave, *Oil Gas J.*, March 27, 1978.
(2) Gulf Universities Research Consortium, "Chemicals for Microemulsion Flooding in EOR," GURC Report No. 159, February 15, 1977.

ROBERT T. JOHANSEN
U.S. Department of Energy
Bartlesville Energy Technology
  Center
Bartlesville, OK 74003
RECEIVED November 14, 1978

ROBERT L. BERG
United States Department of Energy
San Francisco Operations Office
Oakland, CA 94612

Cation Exchange, Surfactant Precipitation, and Adsorption in Micellar Flooding

ROBERT D. WALKER, JR. and W. E. RAY—Department of Chemical
Engineering, University of Florida, Gainesville, FL 32611

M. K. THAM—USDOE Bartlesville Energy Technology Center,
Bartlesville, OK 74003

M. C. LEE—Department of Chemical Engineering, Oklahoma State
University, Stillwater, OK 74074

It is commonly known that the constituents of a micellar slug
may interact in several ways with both the rock and the formation
fluids when injected into a reservoir, and a considerable body of
literature exists (1-8). In spite of this knowledge, however,
it is not yet possible to design a micellar slug for tertiary oil
recovery from basic principles because of the complexity of the
phenomena and inadequate understanding of the processes involved.
The primary objectives of this paper are to present the results of
some experiments on the structure and mineralogy of selected rock
and reservoir core samples, on the interactions within surfactant
solutions and between surfactant solutions and rock, and to
attempt to draw from these observations some conclusions as to the
phenomena and mechanisms involved-especially surfactant loss
processes-as these can affect the maintenance of low interfacial
tension between oil and water.

In the course of attempts to determine adsorption isotherms
of anionic surfactants on selected clays two other phenomena re-
quiring separate investigation were noted, namely, salting-out of
surfactants by NaCl, and surfactant precipitation as calcium or
magnesium salts by multivalent cations displaced from clays.
Each of these and their significance for adsorption measurements
will be dealt with prior to discussion of surfactant adsorption.

Electron Microscopy of Selected Reservoir Core Samples

Scanning electron micrographs of fracture surfaces of Berea
sandstone and two reservoir cores (made available to us by Dr.
F.W. Smith-ARCO Research Laboratories) are shown in Figure 1.
The sample identifications are as follows: 1) Berea sandstone,
Amherst, Lorain County, OH, 2) Glenn Sand, Glenn Pool Field,
Creek County, OK, 3) San Andres Formation, Wasson Field, Yoakum
County, TX. The primary purpose of presenting these is to illus-
trate the geometric and mineralogical heterogeneity of typical
reservoir cores, and the SEM in Figure 1 illustrate these points
quite graphically. The basic sand matrix is clearly visible in

Figure 1.   Electron micrographs of Berea sandstone and selected core samples:
(a) Berea sandstone, representative fracture surface, 102.75×;(b) Berea sandstone,
clay on quartz crystals, 959×;(c) Glenn sand core, representative fracture surface,
123.3×;(d) Glenn sand core, clay crystals on quartz, 3938.75×;(e) San Andres
core, representative fracture surface, 123.3×;(f) San Andres core, clay and dolomite crystals, 993.25×.

the low magnification SEM, while those at higher magnification
reveal the heterogeneity of structure of clays and other
minerals, and their general distribution over the surface of
the sand rather than being concentrated in cementation bridges
between sand grains.

Energy dispersive X-ray analysis (EDXA) is of assistance in
identifying the principal chemical elements in particular
crystals. This information, along with crystal shape, enables
one to identify reasonably well the minerals likely to be contacted by a surfactant slug when injected into a core or a reservoir formation. Thus, the basic sand matrix of these materials
is revealed while the presence of particular clay minerals, such
as kaolinite, can be seen in Berea sandstone and Glenn Sand; dolomite appears to be present in significant amounts in the core
from the San Andres formation.

These SEM, then, show clearly the typical geometric and
mineraological heterogeneity of reservoir rocks. They also give
one an idea of the shapes and sizes of clay and mineral crystals,
and they suggest intimacy of contact between micellar fluid and
clay crystals. Finally, in some cases they suggest the possibility of surfactant precipitation, e.g., dolomite in the San Andres
core sample.

Salting Out And Precipitation of Surfactants By Electrolytes

Reservoir brines and surfactant formulations normally
contain substantial concentrations of electrolytes; one wt.% NaCl
or greater is typical. They may also contain significant concentrations of multivalent cations but one usually attempts to
minimize these because of the low solubility of the sulfonates
of multivalent cations. Owing to the high surfactant concentrations normally used in micellar slugs (typically 5 wt.%), electrolyte effects tend to be masked in experiments with cores. When
one is studying the equilibrium adsorption, however, they cannot
be ignored; indeed, they may be the principal phenomena which are
observed.

Surfactant Salting Out By Sodium Chloride. For the most
part this study has been confined to desalted, deoiled alkylbenzene sulfonates (for procedures of desalting and deoiling see
Ref. 9). Low equivalent weight alkylbenzene sulfonates are so
water-soluble that they are extremely resistant to salting out.
However, when the alkyl chain contains more than about 12 carbons,
salting out becomes increasingly significant. Short chain alcohols are commonly added to micellar solutions to stabilize them
against precipitation, but it should be noted that these concentrated surfactant solutions normally contain about 5 wt.% surfactant and are usually turbid.

In preparing dilute solutions of a series of alkylbenzene
sulfonates for adsorption experiments, it was observed that most
of them became cloudy upon the addition of NaCl, and that a precipitate of salted-out surfactant formed in a considerable number.
Since it was deemed necessary to use clear surfactant solutions
only for adsorption measurements, a more detailed study of the
salting-out phenomenon was undertaken, and some of these results
are presented here.

Aqueous stock solutions of selected anionic surfactants were
prepared; if alcohol was to be added, it was incorporated in the
stock surfactant solution. These surfactants were desalted and
deoiled. Aqueous NaCl solutions were also prepared. Surfactant
(or surfactant/alcohol) and NaCl solutions were mixed in 10 ml
screw-capped test tubes in the proportions necessary to give the
desired final concentration of each constituent. After thorough
mixing, the test tubes were set aside and observed periodically
for clarity and/or precipitation. The results are summarized in
Table I.

In general, cloudiness or precipitation developed almost in-
stantaneously if it occurred at all. The results shown in Table
I indicate that sodium dodecylbenzene sulfonate (SDBS) alone of
the surfactants tested was stable in salt solutions. As little as
0.1 wt. % NaCl caused cloudiness in solutions of all of the other
surfactants tested. The addition of 5 wt.% n-butanol prevented
salting out of sodium pentadecylbenzene sulfonate (SPBS) at both
0.1 and 1.0 wt.% NaCl, but none of the other butanols were effective in preventing salting out of the surfactant. It also seems
worth noting that the 5 wt.% n-butanol solutions of SPBS became
hazy at 1.25 wt.% NaCl and a clear lower phase separated at 3 wt.
% NaCl leaving a hazy upper phase. It appears, then, that salting-
out of surfactant occurs when the surfactant equivalent weight
exceeds 350 and the NaCl concentration exceeds 0.1 wt.%. The
addition of short chain alcohols seems to be effective only for
SPBS (Eq. wt. = 390.5). Although the data are not included in
Table I, it has been noted that 0.1 wt.% solutions of SPBS, TRS
10-410, and Aerosol OT also become cloudly upon the addition of
1.0 wt.% NaCl.

Aside from the significance of the salting-out phenomenon itself, these observations are important for adsorption measurements
in that it appears that the surfactant concentration actually in
solution is less than 0.1 wt.% when appreciable concentrations of
NaCl are present. Not only does the dissolved surfactant concentration appear to be less than about 0.1 wt.% but there is the
effect on the apparent adsorption if the salted out surfactant
partially or completely separates with the clay or other adsorbent
being studied. Complete separation of the salted-out surfactant
leads to large values of apparent adsorption and low equilibrium
surfactant concentrations; negligible separation of the salted-out
surfactant leads to low adsorption and large apparent equilibrium
surfactant concentrations but the actual dissolved surfactant con-

centration may be quite low. It should be noted here that there
is little or no evidence for micelle adsorption of anionic surfactants.

These experiments have made clear that some of our own equilibrium adsorption data are erroneous. It seems possible that
some of the data in the literature may have been affected by
salting-out of the surfactant, and some of these may warrant re-
examination.

Surfactant Precipitation By Multivalent Cations

Precipitation of anionic surfactants by multivalent cations
is well known and it has been studied intensively ([10], [11], [12],[13]).
Powney and Addison ([10]) found that the addition of small amounts
of $CaCl_2$ to dilute sodium dodecyl sulfate solutions caused precipitation of the calcium salt. The addition of small amounts of
n-hexanol was reported to postpone precipitation to higher $CaCl_2$
concentrations, and increasing the surfactant concentration to a
value greater than the CMC was found by Pearson and Lawrence ([13])
to prevent precipitation of the calcium salt of dodecyl sulfate
owing to fixation of calcium ions by the micelles.

More recently Smith ([8]), and Hill and Lake ([14]) studied
cation exchange as it affected the behavior of micellar slugs in
typical reservoir cores. These authors found that cation exchange in cores was quite complex, but that calcium and magnesium
could, for all practical purposes, be treated as a single species.
Moreover, they found that pre-flushing of a core reduced surfactant losses in most cases. Hill and Lake found that surfactant
adsorption in cores was reduced by dissolution of carbonate
minerals and by converting the clays to their sodium form.

Where surfactants were used in these experiments they were
present in relatively large concentration. Hill and Lake, for
example, injected a solution containing 0.046 meq/ml of a surfactant mixture having an average equivalent weight of 410; thus,
the surfactant concentration was about 1.9 wt.%. Since this concentration is far above the CMC for the surfactants involved,
multivalent cations may be bound by the micelles with the result
that calcium sulfonate precipitation does not occur.

In equilibrium adsorption experiments, however, one must
work with much smaller surfactant concentrations and those dilute
surfactant solutions behave differently than concentrated ones.
This was made evident in a dramatic way in two different experiments. In the first case, when a 6 inch column of crushed Berea
sandstone was cation exchanged with 1.0 N NaCl, a slug of solution
containing more than 200 ppm Ca^{+2} (measured by atomic absorption)
and about 0.5 pore volumes wide issued from the column. The
peak Ca^{+2} concentration was about 1500 ppm. When a small amount
of this solution was added to an approximately one wt.% SDBS solution, a copious white precipitate formed. In another experiment, a solution of SDBS in 1.0 wt.% NaCl was added to a sample of

Mississippi montmorillonite in the ratio of 10 ml of liquid per
gram of clay. After equilibration and centrifugation to separate
the solids, it was noted that there were two layers of solids;
the lower one was tan in color and due to the clay; the upper
solid layer was an off-white color and it was later shown to consist mostly of surfactant precipitated as calcium and magnesium
sulfonates.

To evaluate the importance of multivalent cation precipitation of surfactants by multivalent cations present in the formation brine or resulting from cation exchange and dissolution of
minerals such as limestone, dolomite, etc., a very limited study
of the calcium tolerance of selected alkylbenzene sulfonates was
undertaken.

Experimental Procedure. The technique used in these experiments was quite simple: A stock solution of the surfactant was
made and $CaCl_2$ solution in varying concentrations was added, the
final volume being kept constant. If alcohol was to be added, it
was incorporated in the surfactant solution in an amount
sufficient to make the final concentration 3 wt.% alcohol. The
tubes were capped and rotated at one rpm for 24 hours in a thermostat set at 25°C. Experiment at 40°C and 60°C were also conducted. A portion of the liquid was transferred to an absorption
cell and the transmission was measured at a wavelength of 650 nm
(it having been determined previously that the solution and the
precipitate were relatively insensitive to wavelength in that
range). The onset of precipitation was chosen as the Ca^{+2} concentration at which the transmission fell below 98% when compared
to the surfactant solution to which no $CaCl_2$ had been added.

Experimental Results With Crude Surfactants. In general, as
the $CaCl_2$ concentration increased no discernable change in the
transmission occurred until the onset of precipitation; after precipitation was initiated, the transmission decreased rapidly and
precipitation was usually noted unless the surfactant concentration was very small. As the $CaCl_2$ concentration was increased
further, the precipitate began to flocculate and the solution
eventually became clear. The addition of 3 wt.% n-butanol and
2-butanol modified the behavior somewhat but did not prevent precipitation.

The calcium tolerance of a selected group of crude alkylbenzene sulfonates is summarized in Table II. These results,
while relatively imprecise, indicate that the calcium tolerance
of these surfactants in dilute solutions is quite low. In
general, the calcium tolerance decreases as the equivalent weight
increases, and the addition of 3 wt.% n-butanol or 2-butanol does
not appear to improve stability. Finally, the onset of precipitation does not appear to be very sensitive to temperature.

Table I.

Surfactant Salting-Out by Sodium Chloride

Surfactant, 1.0 wt.%		SDBS	SPBS	TRS-10-410	Aerosol OT
Surfactant Equiv. wt.		348.5	390.5	418	444.5
NaCl wt.%	Alcohol 5 wt.%				
0	None	clear	clear	clear	clear
	n-BuOH	not tested	not tested	not tested	not tested
	i-BuOH	not tested	not tested	not tested	not tested
	t-BuOH	not tested	not tested	not tested	not tested
0.1	none	clear	cloudy	cloudy	cloudy
	n-BuOH	not tested	clear	cloudy	cloudy
	i-BuOH	not tested	cloudy	cloudy	cloudy
	t-BuOH	not tested	cloudy	cloudy	cloudy
1.0	none	clear	cloudy	cloudy,ppt.	cloudy,ppt.
	n-BuOH	not tested	clear*	cloudy,ppt.	cloudy,ppt.
	i-BuOH	not tested	cloudy	cloudy,ppt.	cloudy,ppt.
	t-BuOH	not tested	cloudy	cloudy,ppt.	cloudy,ppt.

*becomes cloudy at 1.25 wt.% NaCl; phase separation occurs at about 3 wt.% NaCl.

Table II.

Calcium Tolerance of Crude Alkylbenzene Sulfonates

Surfactant	Temp. °C	Ca^{+2} Necessary to Initiate Precipitation, ppm		
		No Alcohol	3 wt.% n-BuOH	3 wt.% 2-BuOH
SPBS	24	24*	24	20
0.00625 wt.%	30	24*	24	20
	40	24*	24	20
	60	24*	24	20
TRS-10-410	24	24	20	20
0.025 wt.%	30	24	20	20
	40	24	20	20
	60	24	20	20
Petrostep 420	24	15	2	2
0.025 wt.%	30	15	2	2
	40	15	2	2
	60	15	2	2
Petrostep 450	24	15	2	2
0.0125 wt.%	30	12	2	2
	40	10	2	2
	60	5	2	2
Petrostep 465	24	15	2	2
	30	15	2	2
	40	15	2	2
	60	15	2	2

PrecisionL approximately $\pm$ 2 ppm Ca^{+2}

Experimental Results with Purified Surfactants. A similar
series of experiments was carried out with desalted and deoiled
alkylbenzene sulfonates. Aside from the apparently smaller
solubility of the purified surfactants, the results were essentially the same as for the crude surfactants. The calcium ion
concentration necessary to initiate precipitation from SDBS
solutions was about 200 ppm. For purified SPBS it was 10 to 20
ppm, and for purified TRS 10-410 it was about 5 ppm. The onset
of precipitation was relatively insensitive to temperature in the
range 25-60°C, and little affected by the addition of 3 wt.% of
either n-butanol or 2-butanol. It should be noted that precipitation occurred in all cases in spite of the fact that the surfactant concentrations were well above the CMC. In short, fixation of calcium ions by micelles did not prevent surfactant precipitation in these experiments.

Significance For Equilibrium Adsorption Measurements. While
these results appear to have little relevance for micellar slugs,
they are quite relevant for equilibrium adsorption measurements.
They show that precipitation of surfactants can be expected to
occur if the calcium ion concentration exceeds the limit required
to initiate precipitation (SDBS ~ 200 ppm; and TRS 10-410 < 10
ppm). Unless the adsorbent is strongly colored and the surfactant
concentration is substantial, i.e., > 0.1 wt.%, one may experience
difficulty in detecting the presence of precipitated surfactant.
If appreciable precipitation occurs, however, it leads to erroneous adsorption data--as we have noted in several cases. The
problem is obviously more serious with the higher equivalent
weight surfactants (Eq. wt. > 400) and these are the surfactants
of greatest interest for improved oil recovery by micellar flooding.
      It should be noted that the surfactants used in these exper-
iments are commercial products; they are complex mixtures of many
compounds and the equivalent weights represent averages. The influence of the mixture is not shown by these experiments; it is
being investigated now and will be reported in a later publication.

The Adsorption of Anionic Surfactants on Clays and Related Material

      The adsorption of surfactants on reservoir sands and clays
has been known from early studies on surfactant flooding, and
from time to time the results of investigations dealing specifi-
cally with adsorption losses have been reported (1-7, 15,16,17).
These have made clear that adsorption losses and selective adsorption are significant factors in determining the efficiency and
economic feasibility of surfactant flooding for enhanced oil
recovery. However, the mechanism of adsorption is not well understood and several other aspects of the process indicate the need
for additional information and understanding. This investigation

was undertaken with the particular objectives of elucidating the mechanism and identifying the principal effects of system parameters.

Experimental: Materials and Procedure. The equilibrium adsorption of sodium dodecylbenzene sulfonate (SDBS), and deoiled TRS 10-410 (a commercial petroleum sulfonate with an equivalent weight of 418) on silica gel (Davison Grade 62), and crushed Berea sandstone was measured at 30°C at two brine concentrations (0 and 1 wt.% NaCl).

A weighed amount of adsorbent which had been dried overnight at 110°C was transferred to a 15 ml screw-capped test tube, 10 ml of surfactant solution containing either no NaCl or one wt.% NaCl were added, and an amount of deionized water sufficient to saturate the adsorbent (determined in a separate experiment in which adsorbent is equilibrated with water or brine at the adsorption temperature) added. The test tube was covered with polyvinyl chloride film, the cap screwed on, and the tubes mounted on a drum which rotates at one rpm at the adsorption temperature. After twenty-four hours rotation (previous experiments had shown that equilibrium was reached in 16 hrs. or less), the tubes were removed, uncapped, and centrifuged to give a clear supernatant liquid. The liquid was sampled and analyzed for surfactant content by complexing the surfactant with methylene blue in an acidified sodium sulfate solution, extracting the complex into chloroform, and measuring the absorbance of the chloroform solution at the absorption maximum of the complex. The equilibrium or residual concentration of solution was calculated from a material balance. The precision of the measurements appears to be ±2%, and the accuracy appears to be about ±5%.

Adsorption on Silica Gel. The adsorption isotherms of sodium dodecylbenzene sulfonate and TRS 10-410 on silica gel at 30°C and pH =5.8 are shown in Figure 2 for zero and one wt. % NaCl. Although the equivalent weights of these surfactants differ substantially (SDBS = 348; TRS-10-410=418) the isotherms are very similar in shape: there is a concave toe, a shoulder, and a long flat plateau in each case. The addition of one wt.% NaCl to the solution results in a sharp reduction in the adsorption plateau (or saturation level) for SDBS (one wt.% NaCl causes salting-out of TRS-10-410, see Table I, so no adsorption isotherm was measured for TRS-10-410 and one wt % NaCl).

The concave shape of the adsorption isotherms at low surfactant concentrations indicates that the presence of some adsorbed surfactant makes easier the adsorption of additional surfactant, and this process continues up to saturation at which point the isotherm breaks over sharply to the flat plateau. In the case of SDBS the critical micelle concentration is known to be 0.056 wt.% or 1.61 x 10^{-3} moles/l ($\underline{18},\underline{19}$). Thus, the shoulder of the adsorption isotherm occurs at or near the CMC. The con-

stancy of the amount of SDBS adsorbed on silica gel as the
micelle size and concentration increases is evidence that micelle
adsorption does not occur. This result is to be expected since
both the silica surface and the micelles are negatively charged
and strong coulombic repulsion exists between them. The isotherm
for TRS-10-410 has the same shape as those for SDBS. However, the
shoulder occurs at lower surfactant concentration and the adsorption plateau is lower than the SDBS isotherms. This is consistent with a smaller CMC for TRS-10-410.

The CMC of TRS-10-410 and its NaCl concentration dependence
are not known with the precision of that for SDBS, but the CMC is
probably somewhat less than 0.5 wt.% ($1.2x10^{-3}$ moles/l). From
Figure 2, one can observe that the shoulder of the adsorption
isotherm occurs near the CMC, and that there is no evidence of
micelle adsorption on silica gel.

The sharp reduction in the adsorption saturation level when
one wt.% NaCl is added to the solution seems to be associated
principally with the effect of electrolyte on the structure of
the electrical double layer and on the influence of NaCl on the
CMC of the surfactants. The CMC decreases as the ionic strength
of the solution increases and this has the effect of reducing the
maximum surfactant monomer concentration.

Adsorption on Berea Sandstone. Berea sandstone was reported
by Malmberg and Smith (20) to consist of approximately 91 wt.%
sand and 9 wt. % clay. The adsorption measurements reported here
are for the crushed sandstone but it should be noted that essentially all of the adsorption occurred on the clay fraction. In a
separate experiment the clay fraction was separated from the sand
and the adsorption of SDBS measured on both fractions. No adsorption on the sand could be detected while strong adsorption on the
clay was found. Moreover, the adsorption on the clay agreed very
well with that found on the original crushed sandstone when con-
verted to a common basis.

The equilibrium adsorption isotherms of SDBS and TRS-10-410
are shown in Figure 3 for zero and one wt.% NaCl. These isotherms
are strikingly different in shape from those obtained with silica
gel though there is some similarity at low residual surfactant
concentrations. In the first place, a maximum in the adsorption
isotherm is observed when the adsorbent is clay. Secondly, the
addition of NaCl results in a significant increase in the amount
of surfactant adsorbed in contrast to the decrease observed when
NaCl was added to systems with silica gel as the adsorbent.

Adsorption maxima have been observed by several investigators
and various hypotheses have been advanced to account for them
(21,22,23,24). The more recent investigations attribute adsorption maxima to micelle exclusion, but this may be too simplistic
an explanation since the concentration of surfactant monomers
does not appear to be greatly diminished by the presence of
micelles and the amount of surfactant adsorbed at high residual

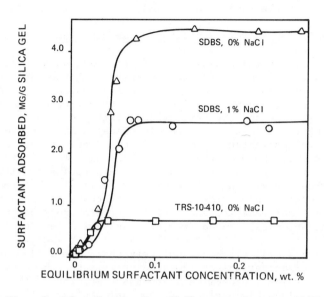

Figure 2.   Adsorption of sodium alkylbenzene sulfonates at 30°C

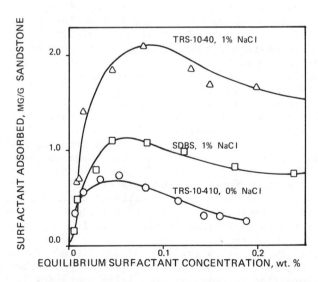

Figure 3.   Adsorption of de-oiled sodium alkylbenzene sulfonates at 30°C

surfactant concentrations may, in some cases, be as little as
20% of that at the adsorption maxima.

The Mechanism of Anionic Surfactant Adsorption on Clay and Silica

Since all of the constituents (surfactant anions and micelles,
as well as silica and clays) are negatively charged at the pH of
the experiments (pH = 6), one must consider why adsorption of the
surfactant should occur at all when electrical repulsion between
the solid surface and the surfactant exists. A possible explanation can be found in the hydrophobic character of the hydrocarbon
"tail" of the surfactant anion and in the tendency for van der
Waals attraction between the hydrocarbon tails to promote adsorption. However, when one calculates the adsorption density and
surface area per molecule, it seems very unlikely that these kinds
of forces could account for adsorption of these surfactants on
silica gel. For example, the minimum surface areas per adsorbed
surfactant anion calculated from the saturation levels of Figure 1
are: SDBS, $4200A^2$ (no NaCl) and $7000A^2$ (1 wt.% NaCl). TRS-10-410,
$31,000A^2$ (no NaCl). It seems unlikely that interaction between
adsorbsed surfactant anions could occur; yet the shape of the toe
of the adsorption isotherm suggests adsorbate interaction.

A more plausible explanation for anionic surfactant adsorption on silica gel is found in the presence of about 0.2 wt.%
alumina in the silica gel. The alumina pzc is about pH=8.0, so it
would be positively charged at the pH of the experiments. If the
alumina is uniformly distributed through the silica, all of the
adsorption could be accounted for provided a close-packed mono-
layer of surfactant is formed on the alumina. This circumstance
would also be consistent with the shape of the toe of the isotherm.
Gaudin and Fuerstenau (25) advanced the idea of hemimicelle formation (two-dimensional micelles on a surface) to account for
similar observations in flotation processes.

In the case of sand in sandstones, the surface area is normally small and the surface charge density is large so that negligible anionic surfactant adsorption is to be expected. The
situation is quite different with the clays commonly present in
sandsontes. The specific surface area is large to very large depending on the mineralogical distribution, and the surface charge
density is less than for silica because the point of zero charge
(pzc) clays is higher than that of silica. Moreover, positive
charges exist at crystal edges and imperfections in clays, and
these can serve as primary adsorption sites for anionic surfactants.

There is little or no evidence of micelle adsorption in
these systems, and, indeed, it is not to be expected owing to the
coulombic repulsion between the negatively charged silica or
clay and the micelles. Surfactant anion adsorption at positive
(cationic) sites followed by further surfactant adsorption induced
by lateral attraction between the hydrocarbon tails to form

hemimicelles appears to be consistent with most of the adsorption data available. There is, however, the question of the orientation of the adsorbed surfactant anions. If the primary adsorption is at positively charged sites, the hydrophobic hydrocarbon tail of the surfactant anion would be presented to the solution. This would tend to produce flocculation and to increase the contact angle at the clay-water interface. Neither of these occurs, at least in our experience. No flocculation of clay has been observed when anionic surfactants are adsorbed; indeed, separation of clay with adsorbed anionic surfactants are adsorbed; indeed, separation of clay with adsorbed anionic surfactants has been somewhat more difficult than when no surfactant was present. Moreover, a few measurements of the contact angle between water and clay with adsorbed anionic surfactant have demonstrated that the clay remains very wettable by water. Contrariwise, the adsorption of a cationic surfactant led to very high contact angles, indicating the development of a hydrophobic water-clay interface.

Berg (26) has found evidence that dimers are the principal species in solution below the CMC, at least when the hydrocarbon chain contains 12 or more carbons. He has proposed that the dimers form by orienting two hydrocarbon tails approximately parallel with a polar head group at each end of the dimer. The adsorption of dimers having the configuration suggested by Berg appears to be consistent with all of the observations made in this laboratory. Moreover, the decrease in CMC with either increase in equivalent weight or with increase in electrolyte concentration is generally consistent with the shifts of the shoulder of the adsorption isotherm.

Conclusion

Typical reservoir rocks are quite complex physically as well as mineralogically. The surface of sand grains available for interactions with injection fluids is a significant fraction of the total surface area; however, the reactivity of sand with anionic surfactants is much less than that between clays and surfactants, so interaction with clays tends to dominate behavior.

Salting-out of alkylbenzene sulfonates from relatively dilute solutions by NaCl has been found to be substantial if the alkyl chain contains more than about 12 carbons. Similarly, the multivalent cation tolerance of alkylbenzene sulfonates in dilute solutions has been found to be small and strongly dependent on equivalent weight. In general, if the equivalent weight exceeds 350 (alkyl chain more than 12 carbons), the calcium tolerance appears to be extremely small. The addition of short-chain alcohols appears to be of limited benefit in preventing either salting-out by NaCl or multivalent cation precipitation.

Anionic surfactants appear to adsorb on silica surfaces, which are negatively charged above a pH of about 2.0, only when

significant amounts of positively charged impurities are present.
Micelle adsorption on silica is not observed, probably owing to
the strong coulombic repulsion between the negatively charged
silica surface and the negatively charged micelles. The
addition of NaCl sharply reduces the adsorption saturation level,
partially as a result of depression of the CMC of the surfactant
at higher ionic strengths but possibly also due to the influence
of electrolyte of the structure of the electrical double layer at
the silica/solution interface. Hemimicelle formation is observed
regardless of the salt concentration.
    Adsorption of anionic surfactants on crushed Berea sandstone
occurs on the clay only and adsorption maxima are observed. The
addition of one wt.% NaCl to the surfactant solution results in
greatly increased adsorption but in no significant change in
the shape of the adsorption isotherm.
    The point of zero charge of the reservoir minerals, their
physical structure, the surfactant equivalent weight and structure, and the structure of the electrical double layer at the
solid/solution interface appear to be major factors determining
the mechanism of adsorption and potential surfactant losses in
surfactant flooding.

Acknowledgement

    The authors wish to express their sincere appreciation to
Energy Research and Development Administration (Grant No. EY-77-
S-05-5341), National Science Foundation-RANN (Grant No. AER 75-
13813) and to the industrial consortium of 21 major oil and chemical companies for their support of the research presented in this
paper. This work is presently supported by USDOE Contract EW-78-
S-19-008.

Abstract

    Scanning electron micrographs of fracture surfaces of Berea
sandstone and representative reservoir cores reveal that the rock
surface is quite heterogeneous as to geometry and mineralogy.
Clay and other minor constituents cover much of the sand rather
than being concentrated in cementation bridges, thus insuring in-
timate contact with the micellar fluid. Alkylbenzene sulfonates
in dilute aqueous solution are very susceptible to salting-out by
NaCl, especially if the alkyl chain contains more than 12 carbons.
Cation exchange capacity measurements show that the CEC of Berea
sandstone can be attributed to its clay fraction and that injection of 1N NaCl into crushed Berea sandstone generates a calcium
ion wave about one pore volume wide with a peak concentration
near 1500 ppm. Measurements of the calcium tolerance of several
sulfonates indicate that precipitation of sulfonate begins at low
calcium concentrations (200 ppm to < 10 ppm depending on sulfonate equivalent weight), and that the onset of precipitation is

not materially affected by added butanol or by sulfonate concentration. Negligible surfactant adsorption is found on sand, but it is appreciable on silica gel. The isotherm on silica gel has a concave toe and becomes constant near the CMC at a level which decreases with increasing NaCl concentration. Sulfonate adsorption on clay is quite different; adsorption maxima are observed and adsorption is increased by NaCl. Adsorbate association (hemimicelle formation) is observed on both silica gel and clay, and some evidence for dimer adsorption is found. Both salting-out and multivalent cation precipitation complicate adsorption measurements.

Literature Cited

1. Gale, W.W., and Sandvik, E.I., Soc. Pet. Eng. J. (1973) 13, 191.
2. Hill, H.J., Reisberg, J., and Stegemeier, G.L., J. Pet. Tech. (1973) 25, 186.
3. Trushenski, S.P., Dauben, D.L., and Parrish, D.R., Soc. Pet. Eng. J. (1974) 14, 633.
4. Somasundaran, P., Healy, T.W., and Fuerstenau, D.W., J. Phys. Chem. (1964) 68, 3562.
5. Bae, J.H., and Petrick, C.B., Soc. Pet. Eng. J. (1977) 17, 353
6. Hurd, B.G., Soc. Pet. Eng. Symposium on Improved Oil Recovery (March 22-24, 1976) Tulsa, OK, SPE Paper No. 5818.
7. Melrose, J.C., Johnson, W.F., George, R.A., and Groenveld, H., Soc. Pet. Eng. 51st Annual Meeting (1976), SPE Paper No. 6049.
8. Smith, F.W., Soc. Pet. Eng. International Symposium on Oilfield and Geothermal Chemistry (1977), SPE Paper No. 6598.
9. ASTM:D2548-69 (Reapproved 1974). "Analysis of Oil-Soluble Petroleum Sodium Sulfonates by Liquid Chromatography", 1976 Annual Book of ASTM Standards, Part 24, Am. Soc. Test. Matls., Philadelphia, PA.
10. Powney, J., and Addison, C.C., Trans. Farad. Soc. (1937) 33, 1253.
11. Shinoda, K., and Ito, K., J. Phys. Chem. (1961) 65, 1499.
12. Cockill, J.M., and Goodman, J.F., Trans. Farad. Soc. (1962), 58, 206.
13. Pearson, J.T., and Lawrence, A.S.C., Trans. Farad. Soc. (1967) 63, 488.
14. Hill, H.J., and Lake, L.W., Soc. Pet. Eng. Fall Meeting (1977), SPE Paper No. 6770.
15. Hower, W.F., Clays and Clay Minerals, (1970) 18, 97.
16. Somasundaran, P., and Hanna, H.S., "Improved Oil Recovery by Surfactant and Polymer Flooding," p.205, D.O. Shah and R.S. Schechter, Eds., Academic Press, New York (1977).
17. Hanna, H.S. Goyal, A., and Somasundaran, P., Paper No. 239, presented at the VIIth International Congress on Surface

Active Substances, Moscow, September 1976.

18. Ludlum, D.B., J. Phys. Chem. (1956) 60, 1240.
19. Tadros, T.F., J. Coll. and Interfac. Sci. (1974) 46, 328.
20. Malmberg, E.W., and Smith, L., "Improved Oil Recovery by Sur-
     factant and Polymer Flooding", p. 275, D.O. Shah and R.S.
     Schechter, Eds., Academic Press, New York, (1977).
21. Meader, A.L., and Fries, B.A., Ind. Eng. Chem. (1952) 44 1536.
22. Eyring, H., and Fava, A., J. Phys. Chem., (1956) 60, 890.
23. Hsiao, L., and Dunning, H.N., J. Phys. Chem. (1955) 59, 362.
24. Mukerjee, P., and Anavil, A., "Adsorption at Interfaces,"
     p. 107, (1975), K. Mittal, Ed., Am. Chem. Soc., Washington,
     D.C.
25. Gaudin, A.M., and Fuerstenau, D.W., (1955) Trans. AIME, 202,
     958.
26. Berg, R.L., (1977) USERDA Bartlesville Energy Research Center,
     Report BERC/TPR-77/3.

RECEIVED November 2, 1978.

The Influence of Surfactant Structure on Low Interfacial Tensions

PETER H. DOE, MAHMOUD M. EL-EMARY, and WILLIAM H. WADE
Department of Chemistry, The University of Texas at Austin, Austin, TX 78712

ROBERT S. SCHECHTER
Department of Petroleum Engineering, The University of Texas at Austin, Austin, Texas 78712

Introduction

The production of a low interfacial tension between an oil and water can provide a mechanism for the recovery of crude oil unobtainable by present techniques. Theoretical considerations and practical results (1,2) suggest that the tension must be brought down near 10^{-3} dyne cm^{-1} to give commercially attractive recoveries. Surfactant systems which can achieve these low tensions have been identified in increasing numbers as research in the area has been stepped up (3-11). The real system which will exist in any final field process will be chemically extremely complex - consisting of a crude oil, a "dirty" surfactant (most probably a petroleum or synthetic sulfonate) and a reservoir brine containing a variety of ionic species. Polymer will also be present.
This complexity makes a study of some of the important variables difficult, because the chemical nature of the system is so badly defined. For this reason it is convenient to use model systems, which it is hoped will adequately duplicate the most essential properties of the complex "real world" situation.
This paper summarizes what has been learned from the study of the simplest possible models and then considers in more detail the application of these results to practical systems.

The Model System

The part of the real world system which is most often dispensed with is the crude oil. From the practical point of view this is desirable in many cases because the dark color of the crude can make the phase behavior of the oil/water/ surfactant system difficult to observe. Crude oils are in any case unpleasant to work with. Hence, Healy and Reed (3-5), in their study of phase behavior, used a synthetic

0-8412-0477-2/79/47-091-017$05.00/0

oil mixture. From a chemical point of view, the crude is
undesirable because of its extreme complexity and the great
difficulty in discovering, even partially, what chemical
species are present (12).

For the work described here, n-alkanes are used as the
oil phase. The results which these yield can be related to
those for other oil structures, including complex mixtures,
by employing the concenpt of equivalent alkane carbon number,
or EACN (13).

The purpose of the present study was to discover the
effect which changing surfactant structure has on the low
tension state. This could only be done if these surfactant
structures were well-defined. For this reason, over sixty
surfactants have been made by sulfonating monoisomeric
hydrocarbons (14-16). In many cases the finished sodium
sulfonate is monoisomeric. In others, up to three isomers
may be present because the sulfonation step is not stereo-
specific. All of these materials are very simple when
compared to the best defined of the commercial synthetic
sulfonates, which contain a minimum of twenty or so species
and a spread of molecular weights.

In the simplest possible case, the systems reported on
below contain only four components--the alkane, water,
sodium chloride and a monoisomeric surfactant. Alcohol
cosurfactants were not usually employed, but their presence
does not appear to influence the results significantly (14).

Experimental Conditions

Studies of low interfacial tension have tended to divide
into two groups--those where a relatively low surfactant
concentration is used, typically 0.2 wt.% or less (6,8-16),
and those where the surfactant concentration is much higher,
above 1% (3-5). There are differences in the way the two
concentration regimes can be handled experimentally.

When the surfactant concentration is high, the potential
for solubilization of either the oil or aqueous phase into
the other is correspondingly high. For this reason, prior
equilibration of the two phases is essential before their
interfacial tension is measured.

With low concentration systems, solubilization effects
are small and preequilibration can be avoided. All interfacial tensions measured for this study were obtained using
the spinning drop technique (17) and a small oil droplet was
simply injected into a tube containing the surfactant formulation without previously contacting these two phases.
Obviously, solubilization phenomena still occur in the low
concentration systems, but dramatic effects, such as third
phase formation or the dissolution of the oil droplet are
not observed.

Recent work ($\underline{18}$) has shown that there are no substantial differences between low tension behavior in the high and low surfactant concentration regimes. This is suggestive of there being a "third phase" present at the interface in the low tension region at low surfactant concentration, as can be experimentally demonstrated for the high concentration, as can be experimentally demonstrated for the high concentrations ($\underline{3}$-$\underline{5}$,$\underline{18}$). This extra phase may simply be too small in quantity to be visible in the low concentration case.

The measurements reported here are confined to a temperature of $26^{\circ}C$. There is no evidence to date concerning the possible influence of temperature on the conclusions reported.

The Alkane Scan

One of the first things that was discovered about the low tension state was that it is very sensitive to changes in system variables. If a low tension is achieved with a given oil/surfactant pair, it is quickly lost if a variable such as salinity, cosurfactant concentration or surfactant molecular weight is changed ($\underline{3}$-$\underline{7}$).

If interfacial tensions are measured against all or part of the homologous series of liquid alkanes ($\underline{8}$-$\underline{10}$), the effect of these same changes is to change the alkane which gives the lowest tension. If the interfacial tensions which a surfactant gives against each alkane are plotted vs. alkane carbon number, as in figure 1, we call the resulting curve the alkane scan of that formulation. It is used to define two variables by interpolation from the experimental curve; n_{min}, the alkane carbon number for minimum tension and γ_{min}, the minimum interfacial tension.

The low interfacial tension properties of pure surfactants will be discussed in terms of these two quantities. The effects of system variables on n_{min} are easier to study and to understand, although ultimately less important, so these will be briefly summarized before discussing effects on γ_{min}.

Effect of Variables on n_{min}

These trends were first observed for commercial surfactants ($\underline{8}$-$\underline{10}$). In most respects pure and complex surfactants are similar. Both types show alkane scans with pronounced minima. In both cases n_{min} is increased by increasing surfactant molecular weight (subject to structural modifications, e.g. ref. $\underline{11}$), by increasing NaCl concentration or by decreasing temperature ($\underline{19}$). Addition of alcohol cosurfactants produces similar n_{min} changes in both cases

and the EACNs of oils measured against all types of
surfactants are the same.

There are some differences between pure and commercial
surfactants which should be pointed out for completeness,
although none of them has any bearing on what follows.
First, n_{min} for a commercial surfactant decreases with increasing surfactant concentration. The degree to which this
occurs depends on the complexity of the mixture and with
single surfactant species it is never observed ([11],[19]).
Second, n_{min} changes with time for petroleum sulfonate systems
([20]). This aging effect is not seen with pure materials.
Both of the above effects are probably related to changes in
the relative abundances of various surfactant species present
as monomer in the solution, arising from micelle-monomer
equilibrium. In a single surfactant system, the monomer is
always 100% of a particular species and hence no n_{min} changes
are observed.

Last, the relationship between n_{min} and "mole fraction"
when two commercial surfactants are mixed is always linear
or very nearly so ([10],[21]). This is never true for a mixture
of two pure surfactants and a variety of different-shaped
curves has been seen. Most often, the n_{min} value of the
mixture is biased towards that of the higher n_{min} component,
which is usually that of the higher molecular weight component.
This presumably reflects high interfacial activity for that
surfactant. When four or more surfactants are mixed, a near
linear n_{min}/mole fraction relationship is once again observed.

The n_{min} idea is useful in describing systematically
the properties of the low tension state ([19]). It has
similar status to the "optimal salinity" favored by other
authors ([3-5]). The one considers the oil required to give
the best low tension when all other variables are held
constant, the other looks for the salinity required for low
tension against a given oil. Similar information about
surfactant behavior is yielded in both cases.

Both approaches, however, are limited in that they do
not directly address the problem of the magnitude of the
minimum tension. If the question is asked--given an oil and
a surfactant, what conditions give the lowest interfacial
tension?--both approaches can quickly give very close to the
right answer. However, if we ask--given an oil, what
surfactant, regardless of other conditions, can give the
lowest tension?--they tell us nothing.

This question can be answered by considering the effect
of system variables on γ_{min}.

Effect of Variables on γ_{min}

This is a fairly difficult thing to study. It has already been pointed out that changing any system variable will change n_{min}. There is always can accompanying change in γ_{min}. However, it is not true to assume that it is the change in the variable which has altered γ_{min}. Consideration of a large amount of data (14-16) has shown beyond reasonable doubt that γ_{min} is linked very closely to n_{min}, but only somewhat loosely to variables such as salinity or surfactant molecular weight.

To understand this a little better, consider the following series of experiments. Take a surfactant formulation whose alkane scan gives a minimum at octane (n_{min}=8). Lower the salinity and decrease n_{min} to 6, so that the lowest tension occurs at hexane. The minimum tension at octane will in general not be the same as the minimum tension at hexane. In other words, γ_{min} for the two scans is different.

Now add sufficient alcohol, say isopentanol, to the formulation to shift the minimum back to octane. It is an experimental observation that the two minimum tensions against octane will be very similar. Decrease the salinity once more, shift n_{min} back to 6, and the two minimum tensions against hexane will be very close. Keeping everything else the same, switch to a surfactant formulation containing surfactant of the same structural type but of a higher molecular weight. The minimum can be shifted back to octane and all three γ_{min} values at octane will be close together. In almost all cases, they will differ substantially from the two γ_{min} values for hexane.

This is a somewhat idealized description. One cannot, in practice, shift n_{min} very far for a single surfactant by changing salinity or alcohol concentration without beginning to lose good low tensions. However, ranges of values of these variables can be established within which they have little effect on the γ_{min} value of a surfactant at a particular n_{min}. Outside these ranges, higher tensions are observed.

Experiments like the one just described can best be summarized by extracting from each alkane scan the values of n_{min} and γ_{min}. A plot of n_{min} vs. γ_{min} can then be made and is found to be independent of salinity (within the range from 0.25 wt% to the point where the surfactant precipitates) and surfactant molecular weight (15,16). The shape of the plot is little affected by addition of an alcohol cosurfactant (14) and is the same for mixed and monoisomeric surfactants of the same structure (14-16).

The shape of the n_{min}/γ_{min} plot is very dependent on the

surfactant structure and three types of curves have been
identified ($\underline{14}$-$\underline{16}$). These are sketched in figure 2. In
every case, the minimum tension is very strongly dependent
on the minimum position within the alkane series. This
means that one surfactant may be very good for giving low
tensions against a particular alkane, while another may be
very poor, even though the experimental conditions can be
adjusted in both cases so that the particular alkane gives
lower tensions than any other. In other words, n_{min} for two
surfactants may be identical, but γ_{min} might differ by
orders of magnitude.

Because of the difficulty of shifting n_{min} more than a
limited amount for a single surfactant species, the curves in
figure 2 were derived by employing surfactant mixtures ($\underline{14}$-$\underline{16}$).
The curves are distinguished by the value of n_o, the optimum
alkane carbon number. This is the alkane against which the
lowest γ_{min} is observed and it varies with surfactant
structure. The need to use surfactant mixtures in defining
the curves means that they can only be resolved, so far, into
the three types shown in figure 2. The curves have been
called preference curves ($\underline{15}$), since they indicate that a
surfactant prefers alkanes of carbon number near n_o to give
very low tensions.

The surfactants which were studied in obtaining these
alkane preference curves were all alkylbenzene sulfonates of
various structures. The preference group to which a
surfactant belongs can be decided if the structure of the
hydrocarbon which was sulfonated is known. Consider the
examples of figure 3. Compound a) is a linear alkylbenzene
sulfonate ($\underline{11}$,$\underline{14}$). There is only one alkyl group and the
SO_3 group enters during sulfonation exclusively in the para
position ($\underline{14}$). Compounds b) and c) have two alkyl groups.
The SO_3 group will enter predominately, but not entirely, in
the least sterically hindered position, as shown ($\underline{15}$). By
similar reasoning, always assuming that the sulfonate group
enters at the position of minimum steric hindrance, the
predominant isomer present in any finished sulfonate can be
decided ($\underline{16}$).

Now, the longest alkyl group on the molecule is regarded
as the major hydrophobic group. Any chain ortho to the SO_3
group is called the interfering group, since it can interfere
with sulfonate-water interactions when the molecule is used
as a surfactant. It has been shown ($\underline{16}$) that the ratio of
the number of carbon atoms in these two groups correlates
strongly with both the surfactant's n_o value and with the
n_{min} values of different surfactants of the same molecular
weight.

In other words, a great deal can be said about the low
tension behavior of any alkylbenzenesulfonate if its molecular
structure is known.

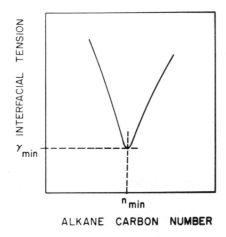

Figure 1. Shape of a typical interfacial tension vs. alkane carbon number plot, defining n_{min} and γ_{min}

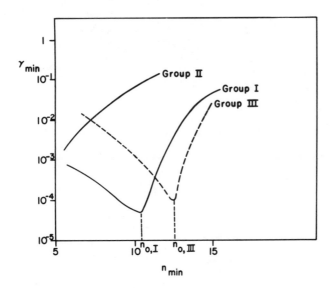

Figure 2. Three types of an alkane preference curve

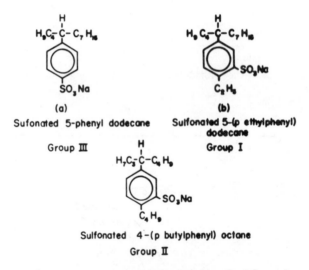

Figure 3. *Sample surfactant structures belonging to each of the preference groups*

a) If it has no interfering group, it is preference group III, n_o = 12.5. Very low tensions are produced only against alkanes in the approximate range decane to tridecane. (i.e. tensions below 10^{-3} dyne cm^{-1}). A long n-alkyl group meta to the SO_3 can act as an interfering group (16).

b) If the interfering group is short relative to the longest group, the surfactant falls into preference group I, n_o = 10. Low tensions are possible against alkanes from pentane to dodecane. (15).

c) If the interfering group is too long, the preference shifts to group II, $n_o < 5$. These surfactants are very mediocre for giving low tensions (15,16).

Thus, in figure 3, compounds a), b) and c) belong respectively to groups III, I and II. The problem of deciding from a surfactant's structure what its preference behavior is has been considered in more detail elsewhere (16).

Crude Oil Systems

The above summary of results has dealt entirely with interfacial tensions against alkanes. It is possible to infer a good deal from this about the probable performance of surfactants when employed against crude oils. The first step in the argument is to introduce the concept of equivalent alkane carbon number (EACN).

An EACN can be defined for any hydrocarbon, hydrocarbon mixture or crude oil by using the following procedure (13). Using surfactants of the same structural type, make up a series of formulations, arranging for each n_{min} value to be different and for these values to span as much of the alkane range as possible. Measure the interfacial tensions of each formulation against the oil whose EACN is desired. Make a plot of interfacial tensions vs. n_{min}. If the curve has a minimum, n_{min} for that point defines the EACN of the oil.

For example, suppose the lowest tension is measured against the surfactant formulation whose n_{min} value is 8. In other words, this formulation gives a lower tension against octane and against the test oil than any of the other formulations. Hence the oil behaves, in a low tension sense, more like octane than any other alkane and its EACN is said to be 8.

EACNs have been determined for several crude oils (13) and fell into a fairly narrow range from 6 to 9. The EACN can be determined in independent ways by using entirely different sets of surfactant formulations. Values checked in this way always agree, confirming that the EACN is an invariant oil property and, in particular, does not vary with surfactant structure (19).

If homologous oil series other than n-alkanes are employed, a similar shape of preference curve for a given

surfactant structure can be demonstrated (14). The
minimum occurs at the same EACN as it does in the alkane
series. Hence n_o represents a preferred EACN among oils of
a similar structure, not merely a preferred alkane.

These pieces of information can be put together to
construct the following thesis. In any series of oils, any
surfactant has a built-in preference for giving low tensions
against oil of certain EACNs. There is a much higher
probability, therefore, of obtaining a low interfacial tension
if n_o for the surfactant is close to the EACN of the oil.

We can go one step further. Suppose crude oils are
regarded as an oil series of varying EACN in the same way
that alkanes are. Then the best low tensions against crudes
should be obtained with surfactants whose n_o values are
closest to the observed 6 to 9 range of crude oils EACNs. In
the terminology used above, this would imply a group I
surfactant. This prediction is based on the assumption that
differences in "structure", i.e. composition, between one
crude oil and another are insignificant in determining their
low tension behavior when compared with differences in EACN.

We have tested this prediction by the following method.
We chose to compare some surfactants of group I--represented
by compound b) of figure 3 and other of similar structure
(15)--with some of group II--similar to compound c) of
figure 3. The crude oils used are lised in table I, together
with their EACNs (13).

TABLE I
EQUIVALENT ALKANE CARBON NUMBER FOR VARIOUS CRUDE OILS

Crude Oil	EACN
Prudhoe Bay field	6.2
West Ranch field	6.6
Wasson field	7.4
Delaware Childers field	7.7
Salt Creek field	7.8
Horseshoe Gallup field	8.2
Big Muddy field	8.5
Bradford field	8.6

A series of surfactant formulations were made up, each of which represented type I or type II preference behavior and each of which had an n_{min} value within the "crude oil" range of 6 to 9. The interfacial tensions of each surfactant were measured against each of the crude oils and several of the lower alkanes.

The results of one of these comparisons for a typical group II surfactant, sulfonated 4(diethylphenyl)nonane, are shown in figure 4. The plots are of interfacial tension vs. alkane carbon number of EACN of the crude. Another comparison is shown in figure 5, this time for a mixture of two group II surfactants, sulfonated 5(p butylphenyl) decane and sulfonated 4(p butylphenyl) octane.

A comparison for a group I surfactant, sulfonated 5(p ethylphenyl) dodecane is in figure 6. These three sets of data are typical of the results. Three things may be noted.

a) The interfacial tensions against crude oils are in general higher than those against alkanes. This is a reflection of the fact that the crudes have a different average structure. A similar difference was noted between alkylbenzenes and alkanes (14).

b) The crude oil results scatter more around the smooth curves drawn through the experimental points than do the alkane results. This is presumably because the alkanes are similar chemical species, whereas the crudes may vary substantially in chemical nature (see comment earlier).

c) The lowest tensions against alkane and crude oil occur at similar EACNs for a given surfactant. This confirms once again that EACN is not dependent on the surfactant used in its determination.

The results of all comparisons which were carried out are summarized in figure 7, where γ_{min} against alkanes is plotted vs. γ_{min} against crudes. There is an obvious correlation between these two. Furthermore, none of the group II surfactants gave an interfacial tensions against a crude oil which was below 10^{-2} dyne cm^{-1}. All of the group I surfactants (solid symbols) did so. All formulations which gave interfacial tensions below 10^{-3} dyne cm^{-1} against alkanes gave values below 10^{-2} dyne cm^{-1} against crude oils.

Clearly, this particular selection of surfactants divides, in their interfacial tension behavior against crude oils, along the lines which were predicted by considering their alkane preferences. This provides support for the idea that estimation of alkane preference by consideration of hydrophobic/interfering group balance (16) can be used as a screening test for deciding whether a surfactant has potential for use in oil recovery. Obviously, the evidence is still limited. In particular, the surfactants compared in the present study have very similar structures and we cannot be

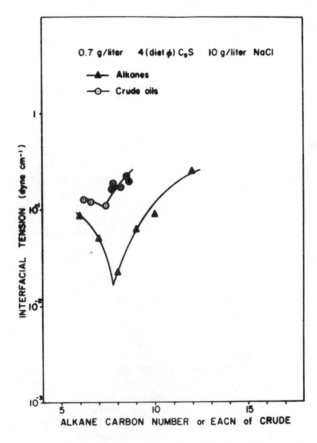

Figure 4.   Comparison of alkane with crude oil interfacial tensions for sulfonated 4(diethylphenyl)nonane, a Group II surfactant

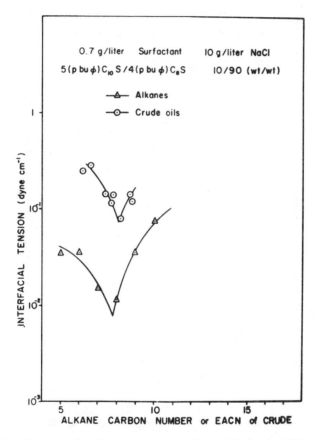

Figure 5.   Alkane/crude oil comparison for a Group II mixture, 90% sulfonated 5(p-butyl-phenyl)decane and 10% sulfonated 4(p-butylphenyl)octane by weight

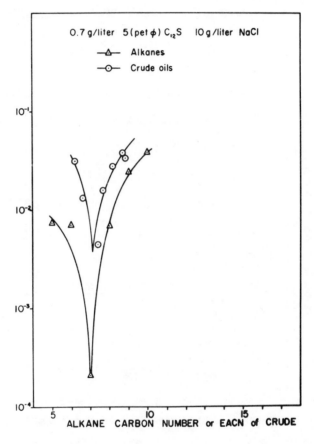

Figure 6.   Comparison of alkane and crude oil tensions for a Group I surfactant,
sulfonated 5(p-ethylphenyl)dodecane

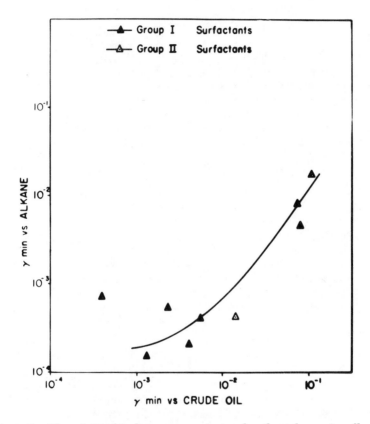

Figure 7. The relationship between γ_{min} *for crude oils and* γ_{min} *for alkanes, showing that a lower tension against an alkane implies a lower tension against a crude for the surfactants examined*

certain, for instance, that all surfactants (or even all
alkylbenzenesulfoantes) will give lower tensions against
alkanes than they will against crude oils. If this is not
the case, knowledge of a surfactant's alkane preference will
be of limited usefulness.

In summary, one may cautiously conclude that a surfactant
of preference group I will be the best type for obtaining
low tensions against most crude oils. Group II surfactants
are almost certain to be ineffective and Group III
surfactants are of intermediate status. Evidence collected
so far (14-16) indicates that these good surfactants will
have a short (methyl or ethyl) group ortho to the sulfonate
group and a relatively long major alkyl chain (10 to 16 alkyl
carbons, or perhaps two shorter groups ortho to each other,
but not to the sulfonate).

Structures like these are present in large numbers in
petroleum sulfonates and synthetic commercial sulfonates.
We may remark that similar alkane preferences can be detected
for these complex commercial mixtures. It is also of
potential importance to note that the EACN of a "live" crude
containing large amounts of dissolved natural gas will be
lower than those of the stock tank oils used in this study.
This would tend to favor surfactants of lower n_o.

Acknowledgements

The authors thank Lex Allen, Charles Bryan, Joel Jeffries
and Paul Kyte for their assistance with interfacial tension
measurements.

This research has been supported by grants from the
National Science Foundation, The Robert A. Welch Foundation,
The Energy, Research and Development Administration and the
following companies: Gulf Research and Development Company,
Pittsburgh, Pennsylvania; Chevron Oil Field Research, LaHabra,
California; Continental Oil Company, Ponca City, Oklahoma;
Shell Oil's Bellaire Research Center, Houston, Texas; Amoco
Production Company, Tulsa, Oklahoma; U. S. Department of
Energy, Bartlesville Energy Technology Center, Bartlesville,
Oklahoma; Sun Oil Company, Richardson, Texas; Witco Chemical
Corporation, Petrolia, Pennsylvania; Mobil Research and
Development Corporation, Dallas, Texas; Atlantic Richfield
Research and Development Department, Dallas, Texas; Exxon
Production Research, Houston, Texas; Stepan Chemical Company,
Northfield, Illinois.

References

1. Taber, J. J., Soc. Petr. Eng. J., 9, 3 (1969).

2. Melrose, J. C. and Bradner, C. F., Journal of Canadian Petroleum Technology, 54-62 (Oct.-Dec. 1974).

3. Healy, R. N. and Reed, R. L., Soc. Petr. Eng. J., 14, 491 (1974).

4. Healy, R. N., Reed, R. L. and Stenmark, D. G., Soc. Petr. Eng. J., 16, 147 (1976).

5. Healy, R. N. and Reed, R. L., Soc. Petr. Eng. J., 17, 129 (1977).

6. Dunlap Wilson, P. M. and Brandner, C. F., J. Colloid Interface Sci., 60, 473 (1977).

7. Wilson, P. M., Murphy, L. C. and Foster, W. R., "The Effects of Sulfonate Molecular Weight and Salt Concentration on the Interfacial Tension of Oil-Brine-Surfactant Systems," Paper SPE 5812. Presented at the SPE Symposium on Improved Oil Recovery, Tulsa, Oklahoma, March 22-24, 1976.

8. Cayias, J. L. Schechter, R. S. and Wade, W. H., J. Colloid Interface Sci., 59, 31 (1977).

9. Cayias, J. L., Schechter, R. S., and Wade, W. H., J. Colloid Interface Sci., 59, 39 (1977).

10. Cash, R. L., et al. p. 1 in "Detergents in the Changing Scene" American Oil Chemists Soc., Champaign, Illinois, (1977).

11. Doe, P. H., Wade, W. H. and Schechter, R. S., J. Colloid Interface Sci., 59, 525 (1977).

12. Rossini, F. D., J. Chem. Education, 37, 554 (1960).

13. Cayias, J. L., Schechter, R. S., Wade, W. H., Soc. Petr. Eng. J., 16, 351 (1976).

14. Doe, P. H., El-Emary, M., Schechter, R. S. and Wade, W. H., "Surfactants for Producing Low Interfacial Tensions, I. Linear Alkyl Benzene Sulfonates," J. Am. Oil Chemists Soc., in press.

15. Doe, P. H., El-Emary, M., Schechter, R. S., and
     Wade, W. H., "Surfactants for Low Tension II. Modified
     Linear Alkylbenzenesulfonates." J. Amer. Oil Chemists
     Soc., to appear.

16. Doe, P. H., El-Emary, M. Wade, W. H. and Schechter,
     R. S., "Surfactants for Low Tension III. Di and Tri
     n-alkylbenzenesulfonates." submitted for publication.

17. Cayias, J. L., Schechter, R. S., and Wade, W. H.,
     ACS Symposium Series No. 8, P. 234 (1975).

18. Wade, W. H., Morgan, J. C., Jacobson, J. K., Salager,
     J. L., and Schechter, R. S., "Interfacial Tension and
     Phase Behavior of Surfactants Systems," Paper SPE 6844,
     presented at the 52nd Annual Fall Meeting of the SPE,
     Denver, Colorado, (October 1977).

19. Morgan, J. C., Schechter, R. S. and Wade, W. H.,
     p. 101 in "Improved Oil Recovery by Surfactant Flooding,"
     (D. O. Shah and R. S. Schechter, eds.) Academic Press,
     N. Y. (1977).

20. Wade, W. H., et al., J. Petr. Technol., 985 (Sept. 1976).

21. Wade, W. H., Morgan, J. C., Jacobson, J. K., and
     Schechter, R. S., Soc. Petr. Eng. J., 17, 122 (1977).

RECEIVED September 18, 1978.

Phase Behavior of a Pure Alkyl Aryl Sulfonate Surfactant

E. I. FRANSES, H. T. DAVIS, W. G. MILLER, and L. E. SCRIVEN

Departments of Chemical Engineering & Materials Science, and of Chemistry,
University of Minnesota, Minneapolis, MN 55455

The phase behavior of surfactants in water and hydrocarbon
is the key to understanding the water- and oil-dissolving power
of certain surfactant systems and the interfacial tension between
the phases that form in these systems (1,2). Ultralow tensions
less than 10μN/m (0.01 dyn/cm) are required by one of the
important mechanisms in various processes for enhancement of
petroleum recovery. Much information is now in the literature
(3,4,5,6), but most of the data are for commercial surfactants
which are complex mixtures of surface-inactive as well as surface-
active components (7).

Recently a University of Texas group synthesized a number of
isomerically pure alkyl aryl sulfonate surfactants (8) which are
structurally similar to components of practically important
petroleum sulfonate surfactant products (8,9). The Texas group
established (9) and we confirmed (10) that ultralow tensions can
be achieved with certain of these pure surfactants, provided the
right combination of surfactant concentration, sodium chloride
concentration, hydrocarbon, temperature and experimental procedure is employed.

Few studies of pure alkyl aryl sulfonate surfactants have
appeared (8,11c,12). The interpretation of phase behavior and
interfacial tension data for pure, single surfactants and well-
defined surfactant mixtures is expected to be easier than for
complex and poorly-defined mixtures. Fundamental studies on pure
surfactants will shed light on the behavior of practically
important surfactant systems which are impure.

Using one of the pure alkyl aryl sulfonates with water,
sodium chloride and decane, we are investigating simultaneously
the phase behavior, the structure of the phases, and the interfacial tensions between them. Ultralow tensions are observed in
this system (10), and it is important to know why they occur,
when they do (13). Our first aim is to establish the equilibrium
phase diagram of surfactant-water-decane as a function of

0-8412-0477-2/79/47-091-035$10.00/0
© 1978 American Chemical Society

temperature and sodium chloride concentration. In this paper we
report the main methods we are applying to phase behavior, some
of which pertain to structure as well, and our results to date.
Interfacial tension measurements will be reported elsewhere (10).
The methods are the following (for additional details see Ref.
14):

Spectroturbidimetry is simply measuring the loss of intensity
of a light beam by absorption and scattering as it passes through
a sample. In terms of transmittance T, the ratio of transmitted
to incident intensity, and total absorbance $A \equiv -\log_{10}T$,

$$\frac{A}{\ell c} = \frac{A_{abs}}{\ell c} + \frac{A_{scat}}{\ell c}$$

where A_{abs} is absorbance due to absorption, A_{scat} is that due
to scattering (also called turbidity), ℓ is the path length (cm)
and c is the concentration (g/cm^3). Since specific absorption,
$A_{abs}/\ell c$, unlike the scattering contribution $A_{scat}/\ell c$, is largely
independent of the sizes and concentrations of particles and
aggregates present (in ordinary solutions it follows Beer's law
and is constant), variations of A/ℓc with respect to concentration, temperature or history reflect changes in sizes and
amount of dispersed material (15). Thus the method can give a
'fingerprint' of a dispersion. It is not sensitive enough,
however, to detect small turbidities arising from ordinary
micelles or comparable molecular aggregates, for which a light
scattering method is required (16). But for this reason
spectroturbidimetry can easily discriminate a fine, non-settling
dispersion from an equilibrium micellar solution.

Polarizing microscopy identifies birefringent phases, for
example liquid crystalline phases (17,18), and reveals dispersed
particles, provided they are larger than about 1µ (ultimate
sensitivity is about 0.3µ).

Vapor sorption measurements yield equilibrium composition
and fugacity or chemical potential; the isopiestic version (19)
is used to determine the uptake of a pure vapor by a nonvolatile
material. This technique determines equilibrium composition of
a phase which cannot be separated quantitatively from the liquid
phase in equilibrium with it. In our application, a nonvolatile
crystalline surfactant specimen S is equilibrated with vapor
of V, which is, in turn, at equilibrium with a system of S and
V consisting of two phases, one rich in S, and one rich in V.
At equilibrium, the Gibbs-Duhem relation guarantees that the
initial specimen of S takes up enough V from the vapor phase
that the chemical potential of S, as well as of V, is the same
as in the biphasic system, and so the composition of the phase
formed by vapor sorption is the same as that of the S-rich phase.
This composition is easily determined by weight measurement. If
the temperature were a triple point, i.e. three phases at

equilibrium, a saturated solution and two surfactant-rich phases, S_1 and S_2, with the same chemical potential of V, then the solid surfactant would absorb V until composition S_2 is reached, whereas the dispersion would consist of phases V and S_1. Thus it has to be verified that the phase produced by vapor sorption is the same as that in the biphasic dispersion.

Carbon-13 Nuclear Magnetic Resonance Spectroscopy (20) gives information about motion — whether fast or slow, isotropic or anisotropic — of specific carbons in the molecular species that make up a sample. Provided a phase has a distinguishing spectrum, its presence (above a limiting mass fraction) can be established, whether or not it is finely dispersed in a second phase. Thus this is a powerful tool for locating phase transitions and probing the phase diagram. It can also be used to help elucidate the microstructure of liquid crystalline phases and micellar aggregates (21).

Conductimetry of aqueous solutions and dispersions of ionic surfactants which are strong electrolytes in monomer form is a sensitive indicator of their state of aggregation, because the equivalent conductivity of even an extremely fine dispersion is much smaller than that of equilibrium micelles which, in turn, is smaller than that of molecularly dispersed ions in water (11b). The conductivity of a micellar solution is due partly to the substantial percentage of the gegenions which are not bound to the micelles, whereas a dispersed phase must retain so much larger a percentage as to leave far fewer gegenions. The percentage of free gegenions in micellar solutions depends on the surfactant head group, micelle size, and concentration. Typical values are 10 to 40% (11b,22). Furthermore, micelles contribute significantly to the conductivity (23), whereas the large particles of a dispersed phase do not.

Ultracentrifugation and ultrafiltration are means of separating dispersed particles of one phase from another and estimating particle size. Ordinary small micelles do not sediment quickly; moreover they re-equilibrate comparatively rapidly after ultracentrifugation. The nature of ultrafiltration membranes used in our application is described in Ref. 24.

APPARATUS AND PROCEDURES

A Cary 15 UV-Visible (200 to 800 nm) Spectrophotometer was used. The cell holder was thermostatted (±0.2°C). Temperatures up to 60°C were measured with a Bailey surface microprobe (±1°C); those between 60°C and 90°C were estimated (±2°C) from the thermostat temperature. Cell temperature was homogeneous within these limits. Total absorbances were measured with an accuracy of 0.003 and a sensitivity of 0.002. They were usually taken at wavelengths between 750 and 350 nm, where scattering is dominant. However, to have sensitivity it was sometimes necessary to work between 350 and 240 nm, where absorption by the surfactant

becomes significant. The latter was evaluated at concentrations
so low that scattering was negligible. Absorbances at the UV
absorption maxima (271.5, 267, 261, 255 and 223 nm) were also
measured and used for analytical purposes after being corrected
for scattering, as determined by extrapolation from the non-
absorbing region of the spectrum. Concentrations found this way
should never be more than 10% above the true value.

Measurements were recorded only after total absorbance
values at several wavelengths were constant for at least two
hours. For biphasic dispersions this was done at several temperatures and a solubility phase boundary was taken to be the
temperature at which the absorbance due to scattering falls
sharply to 0.02 or less. This is the 'synthetic' method for
determining solubilities (25).

Millipore (M) and Nuclepore (N) ultrafiltration membranes
with pore sizes 0.05 to 0.22μ were used to separate dispersed
particles. Alternatively or in alternation a Beckman L3-50
preparative ultracentrifuge with an SW-1 swinging bucket rotor
was used. Initial solutions and dispersions were stirred mag-
netically for at least 12 hours.

A Leitz polarizing microscope equipped with a Nikon photomicrographic attachment was employed with either a slide-
coverslip arrangement or a square capillary specimen container.
Some specimens were taken from equilibrated biphasic systems and
some were produced by vapor sorption of water or decane by pure
surfactant.

For vapor sorption measurements a thermostatted vapor
sorption apparatus (Worden Quartz Products, Inc.) was used (26).
A small pan containing about 100 mg of surfactant was suspended
by a quartz spring, the extension of which was followed by means
of a traveling microscope. The spring was calibrated under
vacuum. Pure solvents (water or decane) were used instead of
saturated solutions in the vapor sorption experiment, because
vapor pressure lowering due to dissolution of surfactant was
estimated to be less than 0.01%, whereas temperature variations
of 0.01°C change the vapor pressure of water by 0.06%. To avoid
weight error because of condensation on the spring or on the
surfactant crystal, the surfactant specimen was kept about 1°C
above the temperature of the solvent. Thus a small systematic
relative error of 5 to 10% in the sorbed amount was introduced.
The steady spring reading, usually attained within 4 to 12 hours
after air evacuation, was taken to be the equilibrium value.
This was confirmed by repetition, partial desorption and re-
equilibration and temperature cycles.

A Varian XLFT-100 Fourier Transform nmr Spectrometer inter-
faced with a Varian 620-L minicomputer with magnetic tape storage
provided high-resolution, proton-decoupled spectra of natural
abundance carbon-13 at 25.2 MHz. For identification of carbon
peaks, chloroform-d solutions of surfactant (solubility about 20
wt%) were prepared. Chloroform-d also served for a deuterium
field lock. Samples of surfactant in water or decane were placed

in 12 mm nmr tubes with a capillary insert containing either D_2O
or acetone-d_6 as lock solvent. The chemical shift references
were TMS (tetramethylsilane) in chloroform-d and acetone-d_6, and
TSP (2,2,3,3, tetradeutero-3 (trimethylsilyl)-propionic acid
sodium salt, by Merck, Sharp and Dohme) in D_2O. Same samples
(Spectra 1,2,5 and 6 shown below) were put in 5 mm NMR tubes and
then an external lock on ^{19}F was employed. The true spectral
width was 5120 Hz (or 5656 Hz), the acquisition time was 0.8s (or
0.727s, respectively), and the sensitivity enhancement time constant was 0.8s. For most spectra, the RF pulse width was 60 μs,
corresponding to a 45° pulse for aliphatic carbons and a smaller
one for aromatics; the exceptions are Spectra 18 and 19, 100 μs,
and Spectrum 5, 113 μs. Ordinarily no pulse delay was used.
The Fourier transform length was always 8192 words. Samples were
spun at about 20 Hz.
    Biphasic solutions were homogenized by bar-stirring after
loading into nmr tubes. Single-phase, surfactant-rich samples
were prepared by equilibrating surfactant crystals inside the
nmr tubes with water or decane vapor (200h for Spectrum 6, 150h
for Spectrum 13) until weight remained constant. The temperature of most of the nmr spectra is 37°C, the temperature to
which the samples are heated by the RF power of the decoupler
when there is no temperature control. However, temperature was
controlled for one-phase systems. The number of transients (T)
and temperatures (°C) are reported below with each spectrum.
    Solution resistances were measured at 1000 Hz with an AC
Wheatstone bridge; an HP 200 AB oscillator provided the input and
a tuned amplifier with a null detector was used to find the
bridge balance (with simultaneous resistance and capacitance
balancing). A homemade cell with Pt electrodes was used,
similar to the one described in Ref. 27a. The cell was calibrated with 0.10N NaCl solutions.

MATERIALS

    The surfactant is the sodium salt of 8-phenyl n-hexadecyl
p-sulfonate (molecular weight 404.6 dalton). One sample was
generously provided by Prof. W. H. Wade of the University of
Texas, Austin. Another batch (Research Sample 8727J) was kindly
provided by Continental Oil Company, Ponca City, Oklahoma.
This 'Conoco sample' was found to contain an impurity insoluble
in chloroform. The chloroform extract was evaporated and dried
in a vacuum oven. The purified Conoco sample gave identical
^{13}C nmr spectra (see Figure 6), the same UV absorption maxima,
and the same polarizing microscope textures as the Texas sample.
The phase behavior and interfacial tensions were similar,
although not always the same (10). Because the first sample was
limited in amount, the Conoco sample was used for some of the
experiments reported here, as indicated by asterisks. Both
appeared to be of high purity in the light of the synthesis

method (8,9) and their ^{13}C nmr spectra, which evidenced no carbon peaks other than those attributable to the presumed structure
(see Figure 6). The origin of the yellowishness of both samples
was not identified. Spectroturbidimetry was conducted at such
low concentrations (<0.7 wt%) that the yellowishness did not
interfere. The surfactant samples were dried at 70°C in a vacuum
oven for at least two days and were stored in a desiccator. The
decane and hexadecane used were 99+% pure Gold Label from Aldrich.
The chloroform was Spectro A.C.S. from Eastman, and the chloroform-d was 99.8% D-pure from Merck, Sharp and Dohme. The sodium
chloride was Certified A.C.S. from Fisher Scientific. Distilled
water was drawn through a Millipore four-stage cartridge system
which reduces conductivity to 0.05 μS cm^{-1}; contact with the
atmosphere and glassware raises conductivity again, however.
Glassware was prewashed, soaked overnight in sulfuric acid-
sodium chromate cleaning mixture, rinsed profusely with plain
distilled water, and finally rinsed with Millipore conductivity
water. Abnormally high conductivities detected subsequently in
some solutions indicated that ions were leached from the glass
over periods of weeks or months.

RESULTS: SURFACTANT-WATER-SODIUM CHLORIDE

Visual and Microscopic Observations

    Stirring 0.7 wt% surfactant in water for several hours pro-
duced a turbid bluish-white system. A specimen under the
microscope and between crossed polarizers revealed numerous
birefringent spherulites 5 to 100μ in diameter, with characteristic Maltese crosses. Similar textures were also observed with
0.3 or 1.0 wt% NaCl present. Figure 1 is a photomicrograph of
typical spherulites together with so-called myelinic figures
(17b), which are caused by interference phenomena.
    Dry surfactant never showed spherulites between crossed
polarizers but did display interference colors in thick samples,
and bright and dark regions in thin samples. Thus the dry surfactant was polycrystalline and structurally different from the
spherulites (see Figure 7). Pressed vigorously between slides,
dry surfactant displayed characteristic ellipses between crossed
polarizers (17a). Evidently surfactant in water absorbed enough
to alter markedly its structure. Evidently too the solubility of
surfactant in water was less than 0.7 wt%.
    Though after stirring the 0.7 wt% system looked homogeneous
to the naked eye for a few days, it then started settling. Sometimes a fairly sharp boundary developed between the upper and the
turbid lower layer. In one such example the lower contained 1
wt% surfactant and the upper, 0.14 wt% surfactant; the overall
content was 0.19 wt%. Under the microscope both layers were
heterogeneous, though there were far more spherulites in the
lower layer than in the upper one. In the absence of salt, many

spherulites with diameter of several microns remained dispersed
for weeks or months; no doubt submicroscopic fragments were even
slower to sediment. The scattering colors of the dispersions
were bluish to grey-bluish, indicating Rayleigh to Rayleigh-
Debye type scattering (15), which is caused by submicroscopic
particles. So even when particles cannot be resolved in the
light microscope the surfactant may not be truly dissolved, but
suspended instead —— and the size, number density, and refractive index contrast of the suspended particles may be too small
to give scattering detectable by eye. Several well-stirred systems (0.10, 0.06, 0.025, 0.014 wt% surfactant) looked transparent
through 2 cm path length under ordinary illumination, unlike
systems with higher concentrations (0.14, 0.22, 0.32, 0.7 wt%),
in which a second phase was confirmed. To test further, water
was layered onto 0.09 wt% dry surfactant without stirring: after
six months only part of the surfactant had dissolved (in contrast,
water layered onto 25 wt% dry sodium dodecyl sulfate dissolved it
all in less than two weeks). Stirred, this composition appeared
to be transparent. It is therefore plain that transparency and
lack of settling in stirred solutions do not guarantee that only
one phase is present. Moreover, path length and illumination
vary with laboratory and observer and are rarely reported. For
these reasons spectroturbidimetry, ultrafiltration and ultracentrifugation were used to detect submicroscopic particles.

Mixtures containing in addition from 0.1 to 10 wt% NaCl were
also examined for particles and scattering. It was found,
qualitatively, that the solubility of surfactant in aqueous salt
solution decreased drastically with increasing salt concentration.
The solubility was less than 0.01 wt% surfactant in the presence
of 0.3 wt% salt. Upper estimates of solubility are summarized in
Table II below.

Order of mixing influenced the state of the dispersion (see
Figure 4 and Table III). In one example, an initially transparent preparation of 0.077 wt% surfactant in water was mixed
with a 3.0 wt% solution of sodium chloride. This yielded a system with 0.30 wt% salt and 0.070 wt% surfactant, which in one
hour looked bluish, then slowly turned turbid-bluish and, in a
few days, developed visible hair-like birefringent particles and
birefringent spherulites. The very same composition when prepared by dissolving the salt, then adding the surfactant, and
finally stirring for 24h, turned out to be much more turbid and
to contain particles most of which settled within days. This
preparation was found to contain a large number of birefringent
spherulites, single or aggregated, of size 5 to 20μ (sizes
depended on the details of stirring). Ultrasonication at 88 kHz
for 10 min made the preparation slightly opalescent, and under
the microscope only a few small (2-3μ) spherulites could be seen.
Spectroturbidimetry showed a ten-fold decrease in turbidity. But
after a few hours, visible particles reappeared, indicating that
the surfactant was not dissolved by ultrasonication but rather

the state of dispersion of the biphasic system was changed.
The supernatant liquids of all <u>unstirred</u> biphasic systems
were transparent. All stirred systems which looked opalescent or
bluish-white precipitated upon standing; in the ultracentrifuge
they yielded a transparent upper layer and a precipitate which
did not redisperse spontaneously even after a month or more.
Thus for these systems, opalescence and turbidity visible over
a 2-cm path length in ordinary illumination, indicated the
presence of a second phase in dispersed form. However, systems
of the same composition looked different not only when mixed in
different orders, but also after heating and recooling, evidently because of dissolution, supersaturation, nucleation and
phase growth which led to different particle size distributions
(<u>15</u>). Apart from these tests systems were ordinarily prepared
by dissolving the salt and then adding the surfactant, at constant temperature, in order to avoid nonequilibrium effects
arising from supersaturation, slow nucleation and sluggish phase
growth.

Certain transparent, one-phase systems after standing for
two months or more became opalescent and threw a liquid
crystalline precipitate: see 0.014 and 0.025 wt% entries in
Table I. Their conductivities were found to have risen markedly,
indicating that ions had been leached from the glass vials, as
discussed in the following section on conductivity. Therefore
the late-forming precipitate was regarded as an artifact and only
the observations and measurements on specimens less than a month
old were reported.

In summary, visual examination and polarizing microscopy
established that at 25°C the surfactant-water phase diagram con-
sists of an optically isotropic phase which is dilute in surfactant, a birefringent phase which is rich in surfactant, and an
extensive two-phase region between. The birefringence and
characteristic microscopic texture mark the surfactant-rich phase
as liquid crystalline (<u>17</u>); moreover it is fluid, as observed in
vapor sorption experiments described below. In aqueous solution
of 2 wt% NaCl or less, dilute suspensions of the liquid
crystalline droplets settled downward into a white or bluish-
white floc layer, whereas in 3 and 10 wt% NaCl, surfactant
particles looked to the naked eye like crystallites of dry
surfactant. However, under the microscope the particles showed
interference colors between crossed polarizers (none between
parallel polarizers), birefringent spherulites and myelinic
figures, indicating that the surfactant-rich phase remained a
<u>liquid</u> crystal even in equilibrium with 10 wt% NaCl solution
despite its <u>solid</u> appearance.

As the proportion of surfactant increased in a biphasic system made grossly homogeneous by magnetic stirring, the viscosity
increased. At 14.5 wt% (no salt) the system was a viscous paste.
At 23 and 30%, it was gel-like; e.g. when turned upside down for

3. FRANSES ET AL.   *Alkyl Aryl Sulfonate Surfactant*   43

Table I. Specific absorbance of surfactant-water mixtures at several wavelengths. A is total absorbance, A_{abs} that due to absorption, $A-A_{abs}$ that due to scattering; ℓ is pathlength and c is concentration

Surfactant concentration wt%	Visual appearance	Absorbance measure	Specific absorbance, cm^2/g Wavelength, λ, nm				
			700	500	436	350	300
0.014	transparent	$A/\ell c$	n.d.[a]	n.d.	n.d.	n.d.	70 ± 15[b]
0.025	transparent	$A/\ell c$	n.d.	n.d.	n.d.	16 ± 8	68 ± 8[b]
0.061	transparent	$\frac{A-A_{abs}}{\ell c}$ [c]	10 ± 3	14 ± 3	23 ± 5	25 ± 10	29 ± 12
0.14	opalescent	$\frac{A-A_{abs}}{\ell c}$ [c]	9 ± 1	24 ± 1	34 ± 1	70 ± 10	110 ± 12
0.70	turbid bluish-white	$\frac{A-A_{abs}}{\ell c}$ [c]	38 ± 1	86 ± 1	112 ± 1	217 ± 10	350 ± 12

(a) absorbance not detectable at 1 cm path length, i.e. less than instrumental sensitivity of spectrophotometer used.

(b) since scattering was not detectable at these concentrations, $A_{abs}/\ell c$ was taken as 16 ± 8 at 350 nm and 70 ± 10 at 300 nm and 0 at 436, 500 and 700 nm.

(c) with $A_{abs}/\ell c$ as explained in b.

a day it did not flow. In none of these was there any sign of
settling even after months, although microscopically they looked
heterogeneous, with birefringent spherulites in a background of
birefringent and non-birefringent patches. The 14.5% system
diluted with salt solution to 13.5 wt% surfactant and 1 wt%
sodium chloride, changed from a translucent, turbid, viscous
paste to a free-flowing, milky, long-lived dispersion. Further
increase of salinity to 10 wt% gave a free-flowing dispersion of
coarser particles which tended to settle upward (the density of
the brine was 1.07 g/cm^3).

Spectroturbidimetry

Systems which appear to be transparent may nevertheless be
biphasic. Representative measurements of absorbances of turbid
as well as transparent systems are reported here to demonstrate
the application of spectroturbidimetry. A system was regarded
as biphasic if one or more of the following criteria were met:

(i)   The wavelength dependence of absorbance was less strong
        than $1/\lambda^4$ in the range 700 > λ > 350 nm. In this range
        absorption by the surfactant is negligible and absorbance
        is due almost exclusively to scattering. This wavelength
        dependence corresponds to Rayleigh-Debye or Mie
        scattering ([15]), which indicates particle sizes greater
        than 0.1μ ([14]), whereas smaller particles give Rayleigh
        scattering, with wavelength dependence $1/\lambda^4$ ([15]).

(ii)   Absorbances though constant over measurement periods of
        hours depended on the order of mixing, age, and thermal
        history of specimens of the same overall composition.

(iii)   The turbidity, i.e. specific absorbance due to scattering,
        $(A-A_{abs})/\ell c$, at 436 nm was larger than 15 cm^2/g. As
        estimated from data on sodium dodecyl sulfate ([16]), this
        corresponds to many thousands of surfactant molecules per
        particle if all of the surfactant is aggregated and thus
        to proportionately more per particle if only a fraction
        of the total surfactant concentration is aggregated into
        particles. In any case this turbidity indicates par-
        ticles considerably larger than even the largest reported
        equilibrium micelles in similar systems ([11b]). When
        absorbances had to be measured at less than 436 nm the
        results were extrapolated to 436 nm in order to apply this
        criterion.

Representative results are shown in Table I and Figures 2
and 3. At 0.014 and 0.025 wt% surfactant no scattering was
detectable and these systems satisfied none of the three
criteria. Furthermore their specific absorbances at 300 nm were
independent of concentration (so were their specific absorbances
at the UV absorption maxima ([14])). It was concluded that

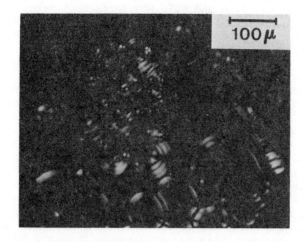

Figure 1. Photomicrograph (×100), with crossed polarizers, of liquid crystals produced by equilibrating dry surfactant crystals with 0.3 wt % aqueous solution of NaCl between two glass slides. The appearance is the same at NaCl concentrations from 0 to 1 wt %.

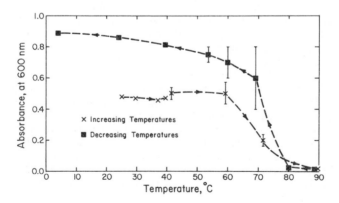

Figure 2. Temperature dependence of total absorbance at 600 nm over 1 cm path length of a biphasic dispersion of 0.70 wt % surfactant in water: bars indicate fluctuations about average absorbance, which was steady for 2 hr or more; arrows indicate direction of temperature change.

absorbances of these dilute systems were due to absorption alone
and could be used to correct for absorption in arriving at the
turbidities $(A-A_{abs})/\ell c$ of the higher concentration samples. The
samples with 0.061, 0.14 and 0.70 wt% surfactant were thus found
to have turbidities at 436 nm larger than 15 cm^2/g. Moreover,
they showed wavelength dependence weaker than $1/\lambda^4$. For example,
at 0.14 wt% the ratio of turbidities at 350 and 700 nm is 8 ± 2 in
comparison with $(700/350)^4 = 16$. So by criteria (i) and (iii)
these systems are biphasic. The differences in turbidity and its
wave-length dependence in these three systems reflected wide
variations in the mean and distribution of particle sizes (15).
It should be noted that the specimen with 0.061 wt% looked quite
transparent.

The dependence of turbidity on temperature and thermal history is shown in Figure 2 for 0.70 wt% surfactant, which is
completely dissolved at 90°C, and in Figure 3 for 0.14 wt%, which
is completely dissolved at 45°C. After a step increase in temperature, average absorbance required a few hours to become
steady, whereas after a step decrease the time required was from
5 to 48h. Occasionally the absorbance would fluctuate around a
steady average value. This is indicated in Figure 2 by bars.
Fluctuations arose from varying scatterer amount in the volume
traversed by the light beam.

Solubilities were determined by applying the foregoing
criteria to a specimen drawn from the uppermost layer of each
system after it had been stirred well (by magnetic stirring bar)
and allowed to settle for two days or more. Table II summarizes
the results: the values listed are the highest concentrations at
which a second phase was not detected by each of the methods
listed. When a second phase had precipitated, concentrations
were determined by UV absorption.

The effect of the order of mixing of surfactant and salt in
water was followed with special attention because of a corresponding effect on the interfacial tension of resulting system
against a drop of decane (10). Figure 4 shows the time dependence
of total absorbance at several wavelengths, of a system which is
biphasic in equilibrium after the salt concentration was raised
from 0 to 0.3 wt%. Turbidity continued to rise for 80h,
especially at lower wavelengths, at which scattering is stronger.
Although the system began settling after 80h, the turbidity continued increasing for a week, and it also continued to increase
after ultrafiltration.

Ultracentrifugation and Ultrafiltration Tests

Tables III, IV, and V show ultraviolet absorption-based
determinations of concentrations after centrifugation (C), ultracentrifugation (UC) and ultrafiltration (UF). The effect of the
order of mixing on size of dispersed particles, as evidenced by
different responses to the same treatment, is plain in Table III.

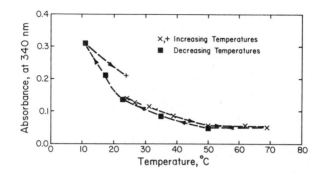

Figure 3. Temperature dependence of total absorbance at 340 nm over 1 cm path length of a biphasic dispersion of 0.14 wt % surfactant in water: each point is the steady value observed for 2 hr or longer; arrows indicate direction of temperature change.

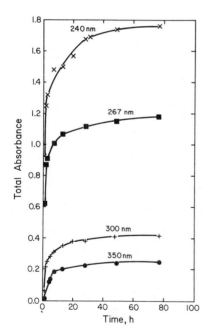

Figure 4. Time dependence of absorbance over 1 cm path length at several wavelengths, of precipitating surfactant system of 0.070 wt % surfactant and 0.30 wt % NaCl in water, prepared by adding 3.0 wt % NaCl solution in time zero to 0.077 wt % surfactant in water (transparent). The initial total absorbance is predominantly by absorption, the increase by scattering from the separating phase.

Table II. Solubility (wt%) of surfactant in water and
salt water as determined by different criteria

Temperature °C	NaCl wt%	Visual[a]	Microscopy[b]	Criteria Spectroturbidimetry[c]	Ultrafiltration[d]
25	0	0.10-0.14	0.10-0.12	0.06	0.05-0.09
30	0	0.1_3		0.10	
45	0	0.2_2		0.14	
60	0	0.3_1		0.23	
90	0	0.7_5		0.70	
25	0.30	<0.008	<0.008	<0.008	0.001-0.002
90	0.30	<0.008		<0.008	
25	3.00	<0.0008	<0.0008	<0.0008	0.0002

(a) Below the concentration cited no particles were visible in liquid
     stirred well and allowed to settle. At 25°C observations were made
     in a sequence of closely spaced concentrations; at higher temperatures
     solubility was defined as the total concentration, at which no visible
     particles remained after heating to the cited temperature.

(b) Above the upper limit, particles were observed at 100X: below the lower
     limit no particles were observed.

(c) Non-Rayleigh scattering, history-dependent scattering, or scattering
     in excess of a certain amount: see text.

(d) Concentration determined by ultraviolet absorbance after filtration
     through Millipore ultrafiltration membrane of nominal pore size 0.05μ:
     See Tables 4 and 5.

Table III. Centrifugation (C) and ultrafiltration (UF) tests of
surfactant solubility (wt%) and particle size of
three-days-old samples mixed in different orders:
in A first the surfactant and then the salt, in B
vice-versa (25°C).

Treatment	A. Precipitation 0.070 wt% surf. 0.30 wt% NaCl	B. Dispersion 0.30 wt% NaCl 0.070 wt% surf.	Remarks
None	0.070	0.070	A scatters less than B
C: 200g, 11h	≤0.065[a]	≤0.03	A scatters more than B
UF: 0.22μ,M[b]	≤0.063[1]		1,2,3 refer to order of filtration through the very same filter
UF: 0.22μ,M	≤0.063[2]	≤0.05[3]	
UF: 0.22μ,M then UF: 0.10μ,M	≤0.047	≤0.045	
C: 200g, 11h; then UF: 0.22μ,M	≤0.045	≤0.020	

(a) Concentrations determined from scattering-corrected absorbance at
ultraviolet absorption maxima; the uncertainty is introduced by the
correction and is estimated to be less than 10% of the value given.

(b) Millipore ultrafiltration membrane (M) of nominal pore size 0.22μ.

Table IV. Ultrafiltration (UF) and ultracentrifugation (UC)
tests of surfactant solubility and particle size
of surfactant-water systems, which were stirred
well and allowed to settle for two days at 25°C.

I)   0.089 wt% surfactant; neither particles nor settling observed; transparent

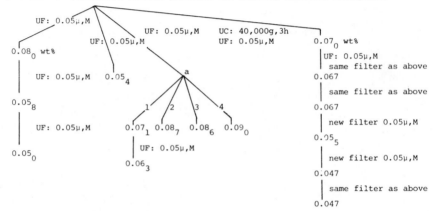

II)   same as I).

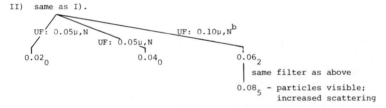

III)   0.135 wt% surfactant; slightly turbid; no particles observed

(a)   Samples 1,2,3,4 correspond to 2 cm^3 aliquots of the same stock preparation
filtered successively, in the order 1 to 4, through the very same filter.

(b)   Nuclepore ultrafiltration membrane (N) of nominal pore size 0.10μ.

Table V. Ultrafiltration (UF) and ultracentrifugation (UC)
tests of surfactant solubility and particle size
of surfactant-water-salt systems; samples were
well stirred and allowed to settle (25°C).

I) 0.30 wt% NaCl   first surfactant, then salt;
    0.084 wt% surfactant   three weeks after preparation.
    supernatant liquid, opalescent, turbid, concentration not determined

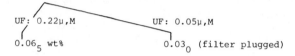

UF: 0.22μ,M   UF: 0.05μ,M

0.06$_5$ wt%   0.03$_0$ (filter plugged)

II) 0.30 wt% NaCl   first salt, then surfactant;
    0.018 wt% surfactant   three days after preparation.
    supernatant liquid, transparent, concentration not determined

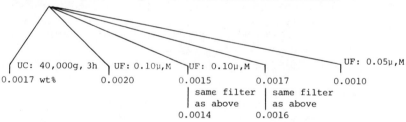

UC: 40,000g, 3h   UF: 0.10μ,M   UF: 0.10μ,M   UF: 0.05μ,M
0.0017 wt%   0.0020   0.0015   0.0017   0.0010
                                  same filter   same filter
                                  as above   as above
                                  0.0014   0.0016

III) 3.00 wt% NaCl   first salt, then surfactant;
     0.0095 wt% surfactant   three days after preparation.
     supernatant liquid, transparent, 0.0008 wt%

                                UF: 0.05μ,M
                           0.0002 wt%

The systems were examined three days after preparation and were gently shaken up before sampling. It is significant that a considerable proportion of the surfactant, which is definitely not in solution (the solubility is less than 0.008% and is probably 0.002%: see Table II) passes through a 0.10µ filter, which is claimed by the manufacturer to retain most particles larger than 0.12µ. In several cases particles were detected in the filtrate by their scattering halos (like dust particles) which suggested that they were larger than the nominal pore size of the Millipore ultrafiltration membrane.

The results shown in Table IV for systems containing no salt revealed that a given filter does not perform reproducibly. Moreover, ultracentrifugation followed by ultrafiltration did not always yield a smaller surfactant concentration in the filtrate than ultrafiltration alone. A new, prewashed (with distilled water) filter always decreased the concentration in the filtrate at first but soon lost its effectiveness; the highest retention was always from the first aliquot. Adsorption in the filter might well have been responsible. Therefore Nuclepore filters (N) of much smaller surface area were also employed. These have straight pores and are much closer to surface filters than are the Millipore filters with their highly interwoven structure and greater thickness (120µ vs. 12µ), which make them depth filters in essence (24). Retention on Nuclepore filters of the same pore size was greater, however, and this argues against retention by molecular adsorption. And in some cases the scattering was stronger after filtration. Despite these problems, the data in Tables IV and V were used to estimate solubilities, with results which are listed in Table II and which do not contradict the values found by other means.

Conductimetry

Equivalent conductivities were first measured of the systems containing 0.014 and 0.025 wt% surfactant (0.35 and 0.62 mM) to shed light on why they threw precipitates months after they had been prepared, tested, and classified as single-phase. The results at these and higher concentrations are given in Figure 5, where data for sodium dodecyl sulfate (28) (which were confirmed to within 10% at low concentrations, 5% at concentrations higher than 5 mM) and for sodium chloride are plotted for comparison. At low enough concentrations that the surfactant is totally dissolved and unassociated, the apparent equivalent conductivities of sodium 8-phenyl-n-hexadecyl-p-sulfonate ought to be about the same as the equivalent conductivities of sodium dodecyl sulfate (SDS), because they have the same cation and this ion contributes the most (around 51/71 = 75%) to the conductivity. It followed that the much larger equivalent conductivities measured in samples six or more months old could be attributed neither to the surfactant alone nor to a surfactant impurity, because they did not correlate to the surfactant concentration. However, for soft glass — the material of the vials employed — contacted by pure water for several months, some 100 ppm of leached solids have been observed (29). Such a concentration of interfering ions can account for the excess conductivity observed (27b). Given the high observed sensitivity of surfactant solubility to sodium ion, it was estimated that the above level of ions could indeed reduce the solubility to less than 0.014 wt% and thus could induce the precipitation mentioned above. Figure 5 also shows that despite the interfering conductivity rise, the equivalent conductivity of the present surfactant falls precipitously with increasing concentration, at 0.7 wt% (17m mol/) reaching 11.2 S cm^{-2} mol^{-1}, which is only 16% of the low concentration value for SDS. This could be accounted for by the concentration of dissolved surfactant — 0.06 to 0.1_2 wt%, according to Table 2 or, in fact, less because of the interfering ions — and partly by ionic contamination. What is most evident from Figure 5 is the marked difference in conductivity between an ordinary micellar solution of surfactant and the aqueous preparations of the present surfactant.

Vapor Sorption

The amount of water sorbed by initially dry surfactant grains was 23 wt% at 25°C (the water reservoir was actually at 24°C, as described above). The difference between this value and the 30 wt% sorbed by surfactant in an nmr tube (Spectrum 6 below) apparently arose in poor temperature control. After water sorption the surfactant grains fused into a continuous film, which indicated the fluid nature of the liquid crystalline material.

Carbon-13 Nuclear Magnetic Resonance Spectroscopy

Carbon-13 spectra for surfactant in chloroform-d are shown in Figure 6. All carbon resonances for the Texas (Spectrum 1) and Conoco (Spectrum 3) samples coincide. Spectrum 2 is for off-resonance proton-decoupling (20b), in which spin-spin coupling should cause peaks of carbons bonded to hydrogen to split — CH_3-peaks into quartets, CH_2-peaks into triplets, CH-peaks into doublets, whereas peaks of non-protonated carbons should remain singlets (20b). The carbon resonances correspond to individual carbons in the surfactant molecule. The assignments are:

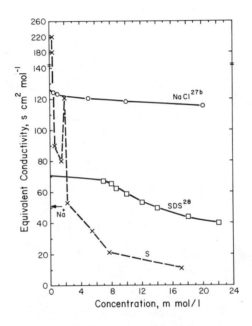

Figure 5. Concentration dependence of equivalent conductivity, at 25°C, of SDS, an ordinary micellar solution, and aged aqueous surfactant (S). One mmol/L of surfactant (S) corresponds to 0.0405 wt %. The critical micelle concentration of SDS is 8 mmol/L. For comparison, equivalent conductivities of sodium chloride and sodium ion, at infinite dilution, are shown.

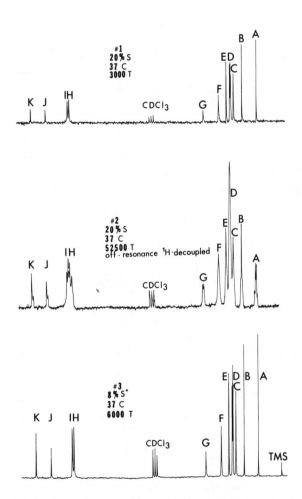

Figure 6. *¹³C, ¹H-decoupled NMR spectra of surfactant in chloroform-D: concentrations are wt % surfactant (S); S* denotes the Conoco sample; temperatures are reported in °C; T stands for number of transients.*

$$\begin{array}{ccccccccccccccccc} A & B & E & D & D & D & C & F & G & F & C & D & D & E & B & A \\ \downarrow & \downarrow & \downarrow & \downarrow & \downarrow & \downarrow & \downarrow & \downarrow & \downarrow & \downarrow & \downarrow & \downarrow & \downarrow & \downarrow & \downarrow & \downarrow \end{array}$$

$$CH_3CH_2CH_2CH_2CH_2CH_2CH_2CH_2CHCH_2CH_2CH_2CH_2CH_2CH_2CH_3$$

Assignments of peaks A through G were made on the basis of published values of chemical shifts of hydrocarbons (30) and comparison with the spectrum of decane (cf. Figure 9, Spectra 12 and 14). Several lines were resolved in the group resonance labeled D. Peaks H and I correspond to protonated benzene carbons, and peaks J and K correspond to non-protonated benzene carbons, by virtue of comparisons with the spectrum of p-isopropyl benzenesulfonic acid, which has the same basic substituents on the benzene ring as the present surfactant. Assignments for this compound were made by using an empirical equation (31) and published values of chemical shifts of the compounds isopropylbenzene and benzenesulfonic acid (30). The assignments agreed with the results of the off-resonance experiment, since A peaks split to a quartet, G, H and I to doublets, and B, E, D appeared to be unresolved triplets (the non-protonated benzene and chloroform-d peaks showed some minor splitting, very likely an experimental artifact). The chemical shifts of the resolved peaks in Spectrum 3 are, in alphabetical order and ppm downfield from internal TMS, as follows: 14.1_3, 22.7_5, 27.7_2, 29.3_6 to 29.9_4, 32.0_7, 36.6_1, 45.9_7, 126.5_5, 127.6_2, 140.6_0 and 149.7_0. Those of chloroform-d were 75.8_7, 77.1_3 and 78.4_1. It is convenient to group A, B, D and E as Class I carbon peaks and all other as Class II carbon peaks.

As shown in Figure 7, only Class I peaks were observed and resolved in Spectrum 4, which is of a dispersion of surfactant in water, 2.3 wt%. According to Table 2, this system contained around 0.06 wt% dissolved surfactant, the remainder being in the form of dispersed liquid crystal. The signal from the dissolved surfactant was too small to be distinguished from the noise, as was evident from Spectrum 5, which is of 0.1 wt% surfactant, for a comparable number of transients. In the middle of Spectrum 5 a negative peak due to the nmr probe carbons appeared, only because of the large number of transients and the extremely low sample signal. The conclusion was that the observed resonances in Spectrum 4 came from the dispersed liquid crystalline phase.

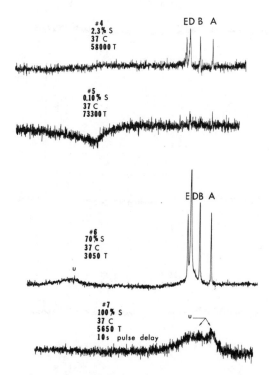

Figure 7. ^{13}C, ^{1}H-decoupled NMR spectra of surfactant–water samples: (#4), dispersion in aqueous solution; (#5), transparent sample; (#6), liquid crystal produced by water vapor sorption by crystal; (#7), dry crystal (u stands for unresolved resonances).

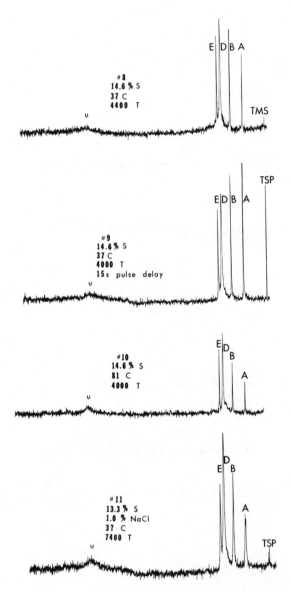

Figure 8.   *¹³C, ¹H-decoupled NMR spectra of surfactant–water samples: relative peak heights are not significant.*

In Spectrum 6, which is for the surfactant-water liquid crystal produced by vapor sorption, only Class I peaks were observed; moreover they were resolved, as in Spectrum 4 of the dispersed liquid crystal (the broad unresolved peak u appeared in 6 but not in 4, probably because of the larger (~6-fold) amount of surfactant in the nmr tube). By this evidence the conclusion was that the dispersed liquid crystal and the liquid crystal produced by vapor sorption were the same.

Spectrum 8, which is for 14.6 wt% surfactant was the same as that of 23 wt% surfactant, of the Conoco sample (not shown), and both were the same as Spectra 4 and 6. This was consistent with both samples being biphasic and only the liquid crystal phase giving an observable signal.

A pulse delay of 15.0s was used to test the sensitivity of the spectra to the pulse parameters: see Spectrum 9. Except for small changes in relative peak heights, which are of little significance in Fourier Transform spectra (20), the spectrum remained unchanged. Changing the pulse width to 20 μs also left the spectrum (not shown) unchanged.

At 81°C only about 0.6 wt% of the 14.6% surfactant was dissolved in water (Table II) and thus most of the surfactant was in the liquid crystalline state. This is evident from Spectrum 10. Finally, the spectrum looked the same when the aqueous phase contained 1.0 wt% NaCl (Spectrum 11).

The sharpness of the Class I peaks suggested that the molecular motion of the corresponding segments of the surfactant molecule in the liquid crystalline phase is fairly fast and probably isotropic. In contrast, the molecular motion of the segments corresponding to Class II peaks is much more restricted, as evidenced by the fact that the carbon peaks were too broad to be observed and resolved. This difference was little affected by temperature, in the range 25 to 81°C, or by salinity in the aqueous phase from 0 to 1 wt%. This pattern of molecular motion stemmed from water absorption by the surfactant, as can be seen by comparing Spectra 6 and 7 of hydrated and dry surfactant, respectively. These data were consistent with the polarizing microscopy evidence of marked structural changes accompanying uptake of water.

RESULTS: SURFACTANT-DECANE

By the spectroturbidimetry and visual observation criteria, the solubility of surfactant in decane was found to be slightly less than 0.04 wt% at 25°C and about the same at temperatures up to about 50°C. Around 50°C, biphasic systems containing 0.046, 0.12 or 7.7 wt% surfactant became clear. The white particles of surfactant-rich phase appeared to melt. A 15.5 wt% sample remained biphasic between 50 and 80°C, in which temperature range the solubility was estimated to be about 9 wt%. The behavior of surfactant in hexadecane was similar: the solubility was about

0.008 wt% at 25°C, changed little up to about 65°C, and rose dramatically above 65°C. These sharp changes in solubility were reversible, though with time lag. When the decane solutions were cooled back down to 25°C, they remained transparent for one to three days; a 0.012 wt% sample in hexadecane remained spectro-turbidimetrically unchanged for more than three days. But after the initial induction period, the surfactant-rich phase began to form visible particles which took a week or more to settle, depending on the overall composition and the thermal history of the sample.

By the isopiestic method the amount of decane taken up by surfactant crystals at 25°C was 19 wt% (decane liquid was held at 24°C). The time required for equilibration was about 12h. Crystals of the Conoco sample equilibrated with decane vapor in an nmr tube for 150h sorbed about 18 wt% decane, in agreement with the result of the isopiestic method. After equilibration the crystallites, though they kept their individuality, looked glossier, seemingly wet. In the polarizing microscope, these crystallites of surfactant-decane phase were birefringent and had a texture similar to that of the dry surfactant. Pressed between slides, they deformed more readily than dry surfactant, which suggested that they were liquid crystalline.

In ^{13}C nmr spectra (see Figure 9) the chemical shifts of the decane-carbon resonances overlap one-by-one with Class I resonances (A, B, D, and E) of the surfactant, provided the following assignments are made:

A B E D D D D E B A
↓ ↓ ↓ ↓ ↓ ↓ ↓ ↓ ↓ ↓
$CH_3CH_2CH_2CH_2CH_2CH_2CH_2CH_2CH_2CH_3$

Thus Class I resonances correspond to either surfactant chains or to decane or to both. Spectrum 12 is from a biphasic 15.5 wt% dispersion at 37°C. No Class II peaks were observed with 1000 transients, nor did the situation change with 10,000 transients (nor did expanding the scale, as in Figure 10, Spectrum 15, make any difference). Thus the concentration of dissolved surfactant was too small to be distinguishable from the noise, in accord with the 0.04 wt% solubility measured at this temperature.

The contribution of dissolved surfactant, whose concentration was only 0.001M, compared to 7M of decane, to the observed Class I peaks must have been negligible. Class II peaks were not observed in Spectrum 13 of the birefringent phase, and Class I peaks were broadened (linewidth about 30 Hz) compared to the peaks in Spectrum 12 (linewidth less than 5 Hz). Therefore it seems quite possible that the dispersed birefringent phase did give Class I peaks in Spectrum 12, but that these peaks, due to either the surfactant or to absorbed decane or to both, merged with those of the decane in the isotropic phase.

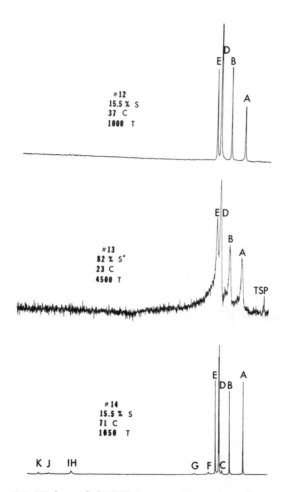

Figure 9. ¹³C, ¹H-decoupled NMR spectra of surfactant–decane samples. The sample with 82 wt % surfactant was produced by equilibrating surfactant crystallites with decane vapor.

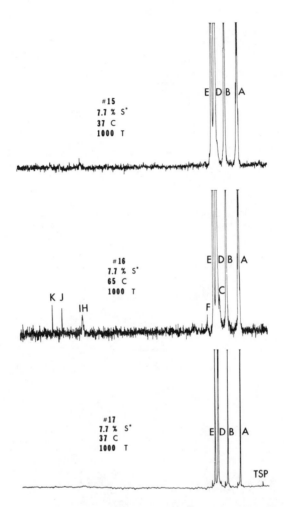

Figure 10. ¹⁰*C,* ¹*H-decoupled NMR spectra over a temperature cycle of 7.7 wt %
surfactant in decane: (#17), 1 hr after cooling; different noise levels are not significant. Vertical scale expanded fivefold over scale of Spectra 12 and 14 (Figure
9), for comparable signal of Class I peaks, to increase sensitivity for Class II peaks.*

The peaks observed in Spectrum 13 indicate hindered molecular motion in the birefringent phase. Further work is needed to resolve which molecules gave the observed signal. That the peaks appear to be superpositions of a narrow peak and a much broader one is tantalizing.

Above 50°C, the solubility of surfactant in decane increased to about 9 wt% and Class II peaks were readily observed at 71°C and 15.5 wt% (Spectrum 14) and at 65°C and 7.7 wt% (Spectrum 16). It was noted that Class I peaks were much more intense than Class II peaks whenever the latter were observed (Spectra 14 and 16), evidently because they came from both surfactant and decane. The latter was present at 30-fold higher molar concentration than the dissolved surfactant.

An hour after the 7.7 wt% sample was cooled back down it looked as clear as at 65°C. However, Spectrum 17 shows that the surfactant molecules were no longer mobile. Evidently they had formed a phase dispersed into units so small — no larger than several hundred Angstroms, probably — that there was no visible scattering. After weeks of standing only a bit of white precipitate appeared, a small fraction of the total surfactant present. It follows that the growth of some of the submicroscopic units to visible, settling size was the slow step in the subsequent precipitation of the surfactant–decane phase.

RESULTS: SURFACTANT-WATER-DECANE

Only a preliminary set of experiments was completed. Polarizing microscopy of two well-mixed systems containing substantial amounts of surfactant, one consisting of 11.9 wt% surfactant-21.4 wt% water-66.7 wt% decane and the other of 21%-71%-8% of the same, revealed birefringent regions and some ellipses resembling focal conics (17a), but no spherulites. Textures such as the one shown in Figure 11 could be altered by rubbing the sample between slide and coverslip. Comparisons with the textures of dry surfactant, of surfactant-water liquid crystal, and of surfactant-decane birefringent phase, indicated that the birefringent material seen in Figure 11 and other cases must have been a surfactant-rich phase which had absorbed both water and decane; however, the samples may have contained more than one phase. On standing, the sample containing 11.9%-21.4%-66.7% separated into a transparent upper layer which was predominantly decane, and a turbid, translucent, viscous lower layer of composition which was estimated to be 17%-33%-50% on the basis of the relative volumes of the lower layer and the upper layer (taken as virtually pure decane). The second system (21%-71%-8%) was gel-like: it did not flow perceptibly when it was turned upside-down for two days, and it showed no sign of settling after more than a month.

Samples prepared with lower surfactant content, yet more than the sum of what would be soluble in the water alone and what

would be soluble in the decane alone, displayed no birefringence.
Figure 12a portrays a sample which originally contained 0.2 wt%
surfactant in water and to which decane was added (the sum of the
separate solubilities fell between 0.07 and 0.1 wt%). After
stirring, the upper layer which formed was milky and viscous; it
did not change noticeably over more than four months; and it
proved to contain a large number of nonbirefringent droplets, of
diameter 10 to 40μ, which were almost close-packed. These
droplets were surely responsible for the intense light scattering
and high viscosity of the layer. At 250X magnification no third
phase could be seen; but since no transparent layers of water or
decane were ever detected, even after four months, it seemed
likely that the surfactant had not actually dissolved completely
in the water and decane, but was incorporated in an undetected
third phase. Moreover, in a sample which originally contained
0.30 wt% NaCl and 0.070 wt% surfactant, and to which decane was
added (the sum of the separate solubilities was about 0.01 wt%),
three phases were observed (Figure 12b); these were a transparent, decane-rich upper layer, an opalescent, water-rich lower
layer, and a floc of white particles. The particles were neither
birefringent nor spherical — in fact they were dimpled — and
they could easily be distinguished in the microscope from the
decane microdroplets which were often entrained during sampling
through the upper phase and which though also nonbirefringent,
were spherical and contrasted differently with the aqueous background. It was concluded that the flocculated white particles
were a third phase rich in surfactant which had absorbed, in
addition to water, sufficient decane to change from more dense to
less dense than the aqueous solution, and to change in structure
from birefringent spherulites to nonbirefringent, nonspherical
particles.

Representative ^{13}C nmr spectra are shown in Figure 13. In
Spectrum 18, which is for the 17 wt% surfactant-33 wt% water-50%
decane sample mentioned above (9:1 molar ratio of decane to
surfactant), only Class I peaks were observed. It was not clear
how many phases were actually present, aside from the birefringent surfactant-water-decane phase. For this phase, which
contains most of the surfactant, absence of Class II peaks indicated hindered motion, as in the surfactant-decane and the
surfactant-water birefringent phases. Class II peaks were not
observed in the spectrum of the 11.9-21.4-66.7% sample (not
shown) which separated into two layers, the upper one being
transparent and predominantly decane. Thus the dissolved amount
of surfactant in the decane which later separated in the presence
of water was also too small to give a detectable signal. In
Spectrum 19, which is for the same sample as above at 71°C, no
Class II peaks were observed either. By comparing Spectra 19 and
14, which is for a sample with no water present, it was inferred
that the presence of water in surfactant-decane samples had one
of the following effects at 71°C: either it decreased the amount

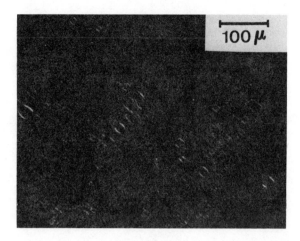

Figure 11.   Photomicrograph (×100), with crossed polarizers, of a gel-like sample containing 21 wt % surfactant, 8% decane, and 71 wt % water, between slide and coverslip

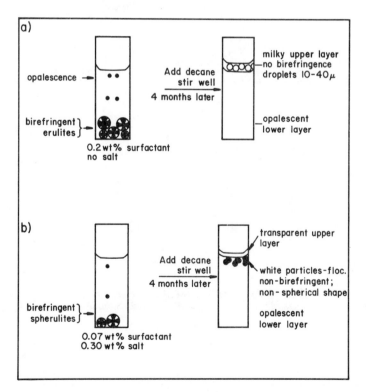

Figure 12.   Schematic of phase behavior and structural changes after mixing a surfactant–brine biphasic dispersion with decane in volume ratio of approximately 5:1

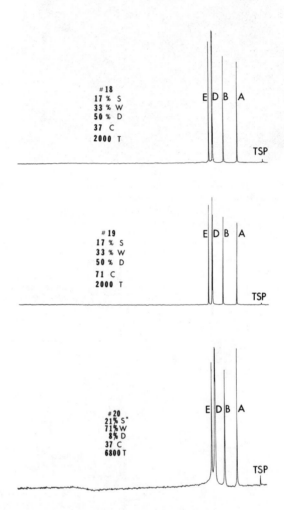

Figure 13. ^{13}C, ^{1}H-decoupled NMR spectra of surfactant (S)–water (W)–decane (D) systems. All samples were translucent to turbid. Sample 20 was gel-like.

of decane-rich phase to an undetectable level, or it altered the solubility of the surfactant in the decane-rich phase, or it simply altered the molecular motion of dissolved surfactant in the decane-rich phase. Spectrum 20, for a different molar ratio (1.1:1) of decane to surfactant, is quite similar to Spectrum 18, indicating the most of the surfactant was not mobile either. Further work is needed to reveal the number of phases present and the molecular motion of surfactant and decane in each of them.

DISCUSSION AND CONCLUSIONS

Surfactant Solubility

The solubility of sodium 8-phenyl-n-hexadecyl-p-sulfonate in water is quite small, 0.06 wt% at 25°C and 0.7 wt% at 90°C. These are the values by spectroturbidimetry, the most reliable technique of those used at these concentrations. The solubility falls rapidly with increasing sodium chloride concentration. The solubility is only 0.0002 wt% in 3 wt% NaCl solution at 25°C (determined by ultrafiltration and analysis by UV absorption). Obviously a common-ion effect is involved in this high sensitivity to Na^+ concentration (32a), but there may be other factors. Thus small amounts of salt impurities which can be leached from a glass vial (29), can cause significant errors in solubility. Solubility can depend on the size of the undissolved phase, though this effect is unlikely to be important for undissolved particle sizes larger than about 1μ when interfacial excess energy is on the order of up to 0.03 J/m^2 (30 erg/cm^2)(32b). Solubility measurements can be confounded by the presence of submicroscopic undissolved particles in stable suspension. In this study the existence was demonstrated of surfactant suspensions in water which are stable for weeks or more, yet to the naked eye are clear and transparent over path lengths of one and two centimeters.

Therefore it is necessary to bring several methods to bear in solubility determinations of surfactants such as the one studied, and it must be recognized that different methods may give somewhat different results. Spectroturbidimetry and nmr spectroscopy help discriminate between dissolved and dispersed surfactant, on the basis of particle size and molecular motion, respectively. Ultrafiltration and ultracentrifugation separations are very useful, but the former suffer the danger of molecular (or micellar) adsorption in the filter and the latter runs the risk of field- or pressure-dependent phase equilibrium. And to interpret the results of either requires spectroturbidimetry and a means of quantitative chemical analysis, such as UV absorption spectroscopy. Fortunately, ordinary equilibrium micelles in solution (33) seem to be unaffected or to re-equilibrate quickly after up to 3h at 40,000g (16) and after passage through a 0.1μ ultrafiltration membrane (34). These filtration results were confirmed in this laboratory by preliminary tests with aqueous sodium dodecyl sulfate below and above the critical micelle concentration (systematic study is under way). Whether micelles of the alkyl aryl sulfonate are present in water at concentrations lower than its solubility is currently being probed by conductimetry. For solubility the method that is usually most reliable, when it is practicable at all, is the spontaneous dissolution of surfactant in the absence of all convection, whether caused by stirring, natural convection, or swelling phenomena (differential swelling may be unavoidable, as for example in the formation of myelinic figures). Though the time required to reach solution equilibrium by diffusion, even in a small volume, can be quite long, this method is being tried too.

The solubility of the surfactant in decane is also quite small at 25°C, about 0.04 wt%, but over a narrow temperature range around 50°C it rises dramatically, as in the Krafft point range of a single-chain surfactant in water (11a). Such a phenomenon with a surfactant in a nonpolar solvent is not uncommon (35). Incidentally, the absence of a Krafft point range for the surfactant in water between 10 and 90°C argues for the absence of micelles in solution. Abrupt change in the slope of such a property as surface tension versus concentration (9) can be due to precipitation of a new phase as well as to onset of appreciable micelle formation, and so does not constitute conclusive evidence for the latter.

Surfactant solubilities in water in the presence of decane, and in decane in the presence of water are important to the ternary phase diagram and are currently under study.

Equilibrium Phases

The saturated aqueous solution of surfactant at 25°C is in equilibrium with a liquid crystalline phase which contains about 25 wt% water. This phase dispersed in solution is the same as the phase formed by water vapor sorption into initially dry surfactant, according to ^{13}C nmr spectroscopy (which virtually eliminates the possibility, mentioned in the Introduction, of a complicating triple point in the two-component system). This hydrated liquid crystal is probably lamellar, to judge by the similarities in texture with lamellar liquid crystals of phospholipids and water (36). It is not uncommon for surfactants for form liquid crystalline phases by absorption of water, or hydrocarbon, or both (37). Moreover the true solubility of many other surfactants (particularly alkyl aryl sulfonates) in water, in salt water, and in hydrocarbon is small, sometimes as small as 0.003 wt% in water, below the Krafft point (38,39). Hence the present finding of liquid crystalline phase in equilibrium with isotropic aqueous solution at surfactant levels above 0.1 wt% may be representative of broad classes of surfactants, including some of interest in connection with surfactant-based chemical flooding processes for oil recovery. The presence of liquid crystalline phases, their intermolecular structure and especially their state of dispersion definitely can affect interfacial tensions and interfacial tension transients (10), and may also influence other factors such as viscosity and the retention of surfactant during flow through a porous medium.

The uptake of water to form liquid crystal has pronounced effects on surfactant molecular motion and intermolecular geometry. The aliphatic carbons near the chain ends move with considerable freedom and give sharp ^{13}C nmr peaks whereas motions in the benzene ring and its proximity on the chain are severely restricted and give peaks too broad to be observed. One consequence is that nmr spectroscopy is a powerful tool for revealing the presence of dispersed liquid crystalline phase even when it is finely subdivided or inextricably incorporated in gel-like material.

Supersaturation and Phase Growth

Supersaturated solutions of the surfactant can be prepared in the usual way by heating a biphasic suspension to a temperature at which all of the surfactant dissolves, and then cooling to the original temperature. They are also generated by mixing saturated or nearly saturated surfactant solution in water with aqueous NaCl solution so that the surfactant solubility is reduced — and the solubility is very sensitive indeed to salinity. In both cases the resulting solution is at first visually clear and transparent and at room temperature remains so for hours or days or, in the most extreme cases, weeks. It may be that nucleation of the partially ordered, liquid crystalline, surfactant-rich phase is slow.

However, within an hour of heating 8 wt% surfactant to dissolution in decane and recooling the nmr spectrum of the dissolved surfactant disappears. Visible particles appear after a few days and begin to settle. Thus even though primary nucleation of the new phase occurs within an hour, the ensuing growth of the particles is extremely slow. Possibly the degree of supersaturation decreases substantially in the first hour and particle growth is by Ostwald ripening i.e. diffusion of surfactant from smaller to larger particles owing to the size-dependence of solubility which is low (32b), or by agglomeration, which can be slow too.

In general, another possibility is that primary nucleation leaves the solution substantially supersaturated yet particle growth is slow; i.e. the system takes a long time to reach equilibrium. It may be possible to resolve these issues by combining spectroturbidimetry to detect changes in state of dispersion with nmr spectroscopy to estimate amounts of dissolved or precipitated surfactant.

Mechanical Dispersion and Settling

The dispersibility of sodium 8-phenyl-n-hexadecyl-p-sulfonate in water is high. Simple agitation by means of a magnetic stirring bar readily converts 0.2 wt% surfactant in water at 25°C into a turbid suspension that persists for days. Dispersions of 1 wt% and higher surfactant proportions in water at 25°C start settling within hours. Given sufficient time the microcrystallites or microdroplets of the dispersed liquid crystalline phase can sediment to produce distinguishable floc layers which seem permanent and in which the overall surfactant concentration is relatively high. Such layers are concentrated dispersions, not equilibrium phases, as the tests by light microscopy, vapor sorption, and ^{13}C nmr spectroscopy prove.

Dispersibility diminishes when the water contains sodium chloride, especially above 0.3 wt% NaCl. Mechanical dispersion of surfactant in decane is also noticeably more difficult. These observations suggest that ionic effects may enter the processes of swelling and subdividing the surfactant into microcrystallites or microdroplets, and that the tendency of the latter to flocculate is influenced by electric double layer phenomena (40).

Stirring systems containing more than 15 wt% surfactant in water produce viscous pastes and gel-like materials which do not settle detectably in months. Dissolving sodium chloride in the aqueous solution can transform a gel-like sample to a low viscosity dispersion.

The explanation for widely discussed but rarely reported effects of the order of mixing alkyl aryl sulfonate surfactant and salt into water (41) appears to be the marked differences in the states of dispersion of liquid crystalline material. At one extreme are systems that continue to develop turbidity for weeks after a transparent surfactant stock preparation is mixed with salt solution. At the other are the mechanical dispersions which settle within hours of being generated by gently stirring surfactant into salt water.

ABSTRACT

Spectroturbidimetry, conductimetry, ultrafiltration, ultracentrifugation, vapor sorption, polarizing microscopy, and nuclear magnetic resonance spectroscopy were used to study phase behavior of pure sodium 8-phenyl-n-hexadecyl-p-sulfonate in water as a function of temperature and sodium chloride concentration, and in decane. The first four techniques gave information on solubility and states of dispersion ranging from visible, settling suspensions to transparent, stably dispersed submicroscopic particles. Surfactant solubility in water was only 0.06 wt% at 25°C, increased 11-fold at 90°C, but decreased 300-fold with 3 wt% salt at 25°C. The surfactant-rich phase in equilibrium with isotropic aqueous solution was shown by polarizing microscopy to be a liquid crystal, probably lamellar; ^{13}C-nmr-aided identification and vapor sorption gave its water content as about 25 wt% at 25°C. Changing the order of mixing of surfactant and salt in water altered the state of dispersion of the liquid crystal present.

Surfactant solubility in decane was 0.04 wt% at 25°C and about 9 wt% at 50°C. The surfactant-rich phase in equilibrium with isotropic decane solution was birefringent. About 20 wt% decane was vapor-sorbed by dry surfactant at 25°C. Preliminary polarizing microscopy and nmr results point toward the existence of liquid crystalline states in surfactant-decane and surfactant-decane-water systems.

ACKNOWLEDGMENTS

We thank Dr. R. M. Riddle and Mr. L. J. Hedlund, not only for obtaining the nmr spectra reported here but also for many helpful suggestions. Mr. T. M. Steele is credited for the conductivity measurements. This reasearch was supported by the National Science Foundation and University of Minnesota.

LITERATURE CITED

1. Healy, R. N., Reed, R. L., and Stenmark, D. C., Soc. Pet. Eng. J. (1976), 16, 147-160, Trans. AIME, Vol. 261, "Multiphase Microemulsion Systems."
2. Healy, R. N., and Reed, R. L., Soc. Pet. Eng. J. (1977), 17, 129-139, "Immiscible Microemulsion Flooding."
3. Taber, J. J., Soc. Pet. Eng. J. (1969), 9, 3-12, "Dynamic and Static Forces Required to Remove a Discontinuous Oil Phase from Porous Media Containing Both Oil and Water."
4. Foster, W. R., J. Pet. Tech. (Feb. 1973), 205-210, Trans. AIME, Vol. 255, "A Low-Tension Waterflooding Process."
5. Melrose, J. C., and Brandner, C. F., J. Can. Pet. Tech. (1974), 13, 54-62, "Role of Capillary Forces in Determining Microscopic Displacement Efficiency for Oil Recovery by Waterflooding."
6. Healy, R. N., Reed, R. L., and Carpenter, C. W., Soc. Pet. Eng. J. (1975), 15, 87-103, "A Laboratory Study of Microemulsion Flooding."
7. Sandvik, E. I., Gale, W. W., and Denekas, M. O., Soc. Pet. Eng. J. (1977), 17, 184-192, "Characterization of Petroleum Sulfonates."
8. Doe, P. H., Schechter, R. S., and Wade, W. H., J. Colloid Interf. Sci. (1977), 59, 525-531, "Alkyl Benzene Sulfonates for Producing Low Interfacial Tensions Between Hydrocarbons and Water."

9.   Wade, W. H., Morgan, J. C., and Schechter, R. S., paper SPE
     6844, presented at the 52nd Annual Fall Technical Conference
     of the Society of Petroleum Engineers of AIME, Denver,
     Colorado (Oct. 1977), "Interfacial Tension and Phase
     Behavior of Surfactant Systems."
10. Puig, J. E., Franses, E. I., Davis, H. T., Miller, W. G.,
     and Scriven, L. E., paper SPE 7055, presented at the SPE-
     AIME Improved Oil Recovery Symposium, Tulsa, Oklahoma,
     April 1978, "On Interfacial Tensions Measured with Alkyl
     Aryl Sulfonate surfactants."
11. Shinoda, K., Nakagawa, T., Tamamushi, B.-I., and Isemura,
     T., "Colloidal Surfactants. Some Physicochemical Proper-
     ties," Academic Press, New York, N.Y., 1963, a) p 7,
     b) pp 25-27, c) p 50.
12. Smith, G. D., Joseph, M. C., and Barden, R. E., J. Colloid
     Interf. Sci. (1976), 56, 395-397, "Some Properties of
     Aqueous Solutions of 2-(n-Tetradecyl)-Naphalenesulfonate
     and 6-(n-Tetradecyl)-Tetralinsulfonate."
13. Anderson, D. R., Bidner, M. S., Davis, H. T., Manning,
     C. D., and Scriven, L. E., paper SPE 5811, presented at
     SPE-AIME Improved Oil Recovery Symposium, Tulsa, Oklahoma,
     March 1976, "Interfacial Tension and Phase Behavior in
     Surfactant-Brine-Oil Systems."
14. Franses, E. I., Ph.D. Thesis, in preparation, University of
     Minnesota, "Phase Behavior and Interfacial Tension Studies
     of Surfactant Systems."
15. Kerker, M., "The Scattering of Light and Other Electro-
     magnetic Radiation, Academic Press, New York, N.Y., 1967,
     pp 338-343 and 396-413.
16. Huisman, H. F., Proc. Kon. Ned. Acad. v. Wet. B. (1964),
     67, pp 367,376,388,407, esp. pp 382,392 and 395, "Light
     Scattering of Solutions of Ionic Detergents."
17. Hartshorne, N. H., and Stuart, A., "Crystals and the
     Polarizing Microscope," E. Arnold Ltd., London (1970),
     pp 504-555, "Liquid Crystals," a) p 513, b) p 554.
18. Rosevear, F. B., Amer. Oil Chem. Soc. J. (1954), 31, 628-
     639, "The Microscopy of the Liquid Crystalline Neat and
     Middle Phases of Soaps and Synthetic Detergents."
19. Glasstone, S., "Textbook of Physical Chemistry," 2nd ed.,
     Van Nostrand, New York, N.Y. (1946), p 632.
20. Broit-Maier, E., and Volter, W., "^{13}C NMR Spectroscopy
     Methods and Applications," Verlag-Chemie, Düsseldorf (1974),
     a) p 23, b) pp 44-49.
21. Roberts, R. T., and Chachaty, C., Chem. Phys. Letters
     (1973), 22, 348-351, "^{13}C Relaxation Measurements of
     Molecular Motion in Micellar Solutions."
22. Vold, M. J., J. Colloid Sci. (1950), 5, 506-513, "The
     Application of the Mass Law to the Aggregation of Colloidal
     Electrolytes."

23. Hartley, G. S., Collie, B., and Samis, C. S., Trans. Farad. Soc. (1936), 32, 795-815, "Transport Numbers of Paraffin-Chain Salts in Aqueous Solution."

24. Hwang, S.-T., and Kammermeyer, K., "Membranes in Separations," in Techniques of Chemistry, Vol. VII, Weissberger, A., and Rossiter, B. W., eds., Wiley-Interscience, New York, N.Y. (1975), pp 425-431.

25. Mulley, B. A., "Solubility in Systems Containing Surface-Active Agents," in Adv. Parm. Sci., (1964), Vol. 1, Bean, H. S., Beckett, A. H., and Carless, J. E., eds., Academic Press, London, pp 86-194, esp. pp 129-132.

26. Rai, J. H., and Miller, W. G., Makromolecules (1972), 5, 45-49, "Vapor Sorption of N,N-Dimethylformamide on Poly (γ-benzyl α,L-glutamate).

27. Shedlovsky, T., and Shedlovsky, L., "Conductometry," in Techniques of Chemistry, Vol. 1, "Physical Methods of Chemistry," Weissberger, A., and Rossiter, B. W., eds., Part IIA, "Electrochemical Methods," Wiley-Interscience, New York, N.Y. (1971), Chap. III, a) p 180, b) pp 168,194, 199.

28. Mysels, K. J., and Otter, R. J., J. Colloid Sci., (1961), 16, 467-473, "Conductivity of Mixed Sodium Decyl and Sodium Dodecyl Sulfates — The Composition of Mixed Micelles."

29. Thiers, R. E., in Methods in Biochemical Analysis, Glick, D., ed., (1957), 5, 273-335, "Contamination in Trace Analysis."

30. Johnson, L. F., and Jankowski, W. C., "Carbon-13 NMR Spectra. A Collection of Assigned, Coded and Indexed Spectra," Wiley, New York, N.Y., (1972), spectra 163,352, 277,428 and 455.

31. Wehrli, F. W., and Wirthlin, T., "Interpretation of Carbon-13 NMR Spectra," Heyden, London (1976), pp 46-47.

32. Kolthoff, I. M., Sandell, E. B., Meehan, E. J., and Bruckenstein, S., "Quantitative Chemical Analysis," McMillan, New York, N.Y. (1969), a) p 126, b) p 210.

33. Tanford, C., "The Hydrophobic Effect. Formation of Micelles and Biological Membranes," Wiley-Interscience, New York, N.Y. (1973), p 36.

34. Parfitt, G. D., and Wood, J. A., Kolloid-Z u.Z. Polymere, (1969), 229, 55-60, "Light Scattering of Sodium Dodecyl Sulfate in Methanol-Water Mixtures."

35. Ekwall, P., and Stenius, P., in Kerker, M., ed., "Surface Chemistry and Colloids," International Review of Science, Physical Chemistry Series Two, Vol. 7, Butterworths, London (1975), pp 215-248, "Aggregation in Surfactant Systems."

36. Papahadjopoulos, D., and Miller, N., Biochim. Biophys. Acta, (1967), 135, 625-638, "Phospholipid Model Membranes. I. Structural Characteristics of Hydrated Liquid Crystals."

37. Winsor, P. A., Chem. Rev. (1968) 68, 1-40, "Binary and
     Multicomponent Solutions of Amphiphilic Compounds.
     Solubilization and the Formation, Structure, and Theoretical
     Significance of Liquid Crystalline Solutions."
38. Ekwall, P., in Advances in Liquid Crystals, (1975), Vol. 1,
     Brown, G. H., ed., Academic Press, New York, N.Y.,
     pp 1-142, esp. p 10, "Composition, Properties and
     Structures of Liquid Crystalline Phases in Systems of
     Amphiphilic Compounds."
39. Gries, W., Fette-Seifen-Anstrichmittel, (1955) 57, 24-33,
     "Über die Beziehungen zwischen der Konstitution und den
     Eigenschaften von Alkyl-benzol Sulfonaten mit jeweils einer
     geraden oder verzweigten Alkylette bis zu 18 Kohlenstoff-
     Atomen I."
40. Adamson, A. W., "Physical Chemistry of Surfaces," 3rd ed.,
     Wiley-Interscience, New York, N.Y. (1976), p 216.
41. Dunlap, P. M., and Foster, W. R., U.S. Patent 3,468,377
     (1969), "Waterflooding Employing Surfactant Solution,"
     P. M. Dunlap-Wilson, personal communication (1977).

RECEIVED August 25, 1978.

Colloidal Properties of Sodium Carboxylates

W. H. BALDWIN[1] and G. W. NEAL[2]

Chemistry Division, Oak Ridge National Laboratory, Oak Ridge, TN 37830

Tall oils are obtained in large quantity from the pulping of soft woods to make paper. Hoyt and Goheen (1) state that in 1964 the yield of fatty and rosin acids exceeded 400,000 tons. Tall oil contains about 94% total acids and 6% alcohols. The fatty acid fraction contains oleic ∿ 24% and linoleic ∿ 23%, of which ∿ 4% is conjugated linoleic of the tall oil. The rosin acid fraction contains abietic acid ∿ 14% and related acids 33% of the total (2). An extensive literature on general aspects of the colloidal chemistry of compounds in this class is available (see, for example, references 3, 4, 5, and 6).

The large supply of tall oils and the well-known surface properties of many of the components have led to several suggestions to use them or their derivatives in micellar flooding (7,8, 9). However, there are, so far as we know, no extensive laboratory investigations underway nor plans to test these possibilities in the field. In view of the contribution tall oils might make to enhanced recovery if they could be used, a survey of interfacial tension properties of aqueous/hydrocarbon systems, similar to those which have become common with the petroleum sulfonate and other surfactants under consideration for micellar floods, seemed worthwhile.

The recent popularity of interfacial tension measurements is largely attributable to the development of the spinning drop approach into a practical technique by Cayias, Schechter, and Wade (10). Gash and Parrish (11) have increased the utility of the method by designing a variation that is capable of making measurements with several samples simultaneously, with the sacrifice of accuracy of the Cayias, Wade, and Schechter device in some ranges of interfacial tensions, because of the constraint of a single rotational velocity.

In the work described here, sodium oleate was selected as

---

[1] Current Address: Rt. 1, Erie, TN

[2] Current Address: Apt. #12, 585 E. Parkway S., Memphis, TN

the model for intensive study of several variables, and other
carboxylates have been included to indicate the effect of structure.

Experimental

The acids: (Figures 1 & 2) 95 and 99% oleic, 95% linoleic,
99% linolenic, 99% elaidic and cholic acids were obtained from
chemical supply houses. Mixed rosin acids were a commercial pro-
duct, Acintol R type S (2). Sodium salts of fatty acids were
formed by electrometric titration of an aqueous-ethanol solution
with sodium hydroxide, evaporated to dryness to remove ethanol
and stored dry or as an aqueous stock solution. A sample of
mixed rosin acids was assayed electrometrically with potassium
hydroxide in benzene-ethanol. A stock solution was then prepared in water from a weighed quantity of Acintol R and the
amount of sodium hydroxide calculated from the assay. Other
materials were reagent grade chemicals.

Interfacial tension measurements were made on one or both of
the devices mentioned (10,11). Spinning of the samples was continued until successive readings were constant; usually 24 hours
was sufficient. All measurements were made at 30°C. Plots of
equivalent alkane number versus interfacial tension were made
with pure n-alkanes - no mixtures were used in results reported
here.

Precision of Measurements. Aliquots from a stock solution
of 0.1 M sodium oleate (five months old) were used to prepare
aqueous test solutions that were 0.01 M in sodium oleate and 0.1
M in sodium chloride pH 9.5. Interfacial tensions were measured
against n-undecane without pre-equilibration. The second solution was made and measured one week after the first and the third
solution two weeks after the first. The results in Table I

Table I
Precision of Interfacial Tension Measurements
0.01 M Sodium Oleate, 0.1 M Sodium Chloride pH 9.5
vs n-Undecane

	Interfacial Tension (dynes/cm)		
	Run I	Run II	Run III
Multi-drop	0.065	0.140	0.062
constant	0.058	0.070	0.054(a)
velocity	0.085	0.068	0.062
	0.062	0.068	0.080(a)
	0.070	0.077	0.068
Avg.	0.068 ± 0.009	0.085 ± 0.022	0.064 ± 0.003(b)
Single-drop	0.098	0.094	0.080
variable			
velocity		Avg. 0.091 ± 0.007	

(a) Air bubbles appeared. Results discarded.
(b) The range is the average deviation from the mean.

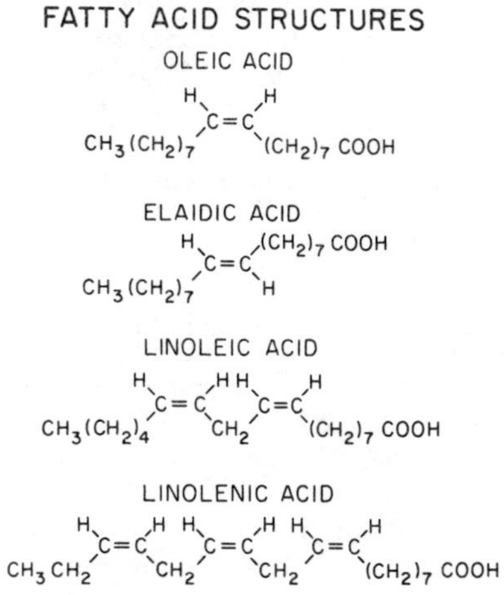

FATTY ACID STRUCTURES

OLEIC ACID

ELAIDIC ACID

LINOLEIC ACID

LINOLENIC ACID

Figure 1. Fatty acid structures

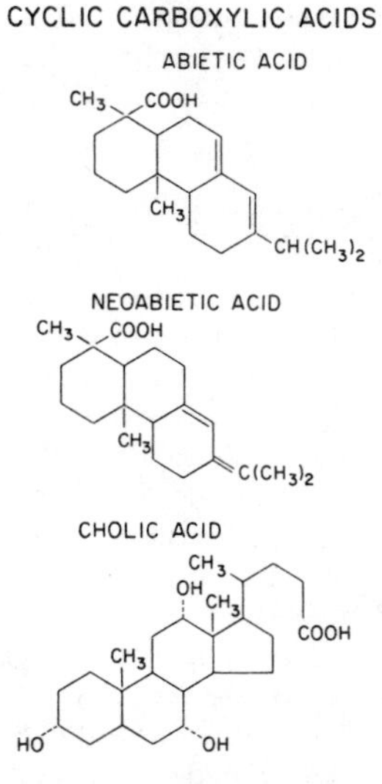

CYCLIC CARBOXYLIC ACIDS

ABIETIC ACID

NEOABIETIC ACID

CHOLIC ACID

Figure 2. Cyclic carboxylic acids

indicate a lower average obtained with the multisample apparatus
than with the single-sample unit, although scatter in the two
sets overlapped. Most points fall within ±0.02 dynes/cm of the
overall average, or within about 30%. Although the uncertainty
appears large, the accuracy is nevertheless adequate to identify
conditions corresponding to the millidyne values favorable for
enhanced oil recovery. These low interfacial tensions typically
occur only for a narrow range of compositions, outside which tensions are orders of magnitude higher than the minimum value.

Results

1. Effect of Surfactant Concentration. Figure 3 compares
results of alkane scans for three concentrations of sodium oleate
at constant sodium chloride concentration and pH. The 0.002 $\underline{M}$
solution is derived from 95% oleic acid, the 0.01 $\underline{M}$ and the 0.1 $\underline{M}$
solutions are derived from 99% oleic acid. Both the magnitude
and the alkane position of minimum interfacial tension ($\underline{n}_m = 11$)
are essentially concentration independent under these conditions.
Wade, et al (12) reported a similar invariance in $\underline{n}_m$ with a pure
alkyl benzene sulfonate, although there was more change in the
minimum value of interfacial tension with the sulfonate concentration than is observed with the carboxylate. The interfacial
tension at $\underline{n}_m$ for 0.01 $\underline{M}$ sodium oleate is in the range of the
values of Table I. Very high interfacial tension (> 10 dynes/cm)
was found at 0.0001 $\underline{M}$ sodium oleate in 0.1 $\underline{M}$ sodium chloride.
This presumably reflects the fact that the surfactant concentra-
tion is almost certainly below the critical micelle concentration (13).

2. Effect of Sodium Chloride Concentration. Figure 4 compares interfacial tensions of several different surfactant concentrations verses $\underline{n}$-undecane in the presence of 0.1 $\underline{M}$ sodium
chloride with values obtained without salt. Salt reduces the
interfacial tension at all surfactant concentrations. Aqueous
potassium oleate has a critical micelle concentration of ∿0.001
$\underline{M}$ (13). It could be inferred from Figure 4 that 0.001 $\underline{M}$ sodium
oleate with no added salt is below the cmc, because of the high
interfacial tension. If so, the much lower interfacial tension
in the presence of 0.1 $\underline{M}$ sodium chloride stems from reduction of
the cmc expected in the presence of added salt (14).
   In Figure 4, interfacial tensions at 0.01 and 0.1 $\underline{M}$ sodium
oleate and 0.1 $\underline{M}$ sodium chloride are higher than in Figure 3.
However, the desired surfactant concentration in Figure 4 was obtained by dissolving the required quantity of sodium oleate and
pH was not controlled. Data from Figure 4 are not directly comparable to Figure 3 for this reason.
   Table II illustrates the effect of varying the sodium chloride concentration on interfacial tension for one surfactant concentration. Between 0.01 and 0.2 $\underline{M}$ sodium chloride there appears
to be a small decrease in interfacial tension. Increasing the

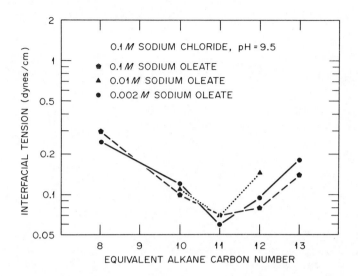

Figure 3.   *Interfacial tension—sodium oleate*

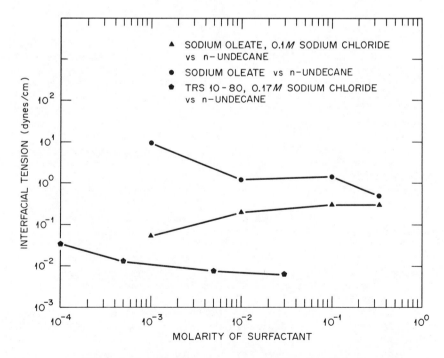

Figure 4.   *Interfacial tension—concentration of surfactant*

sodium chloride concentration much above 0.2 $\underline{M}$ leads to gel formation (15).

Table II

Interfacial Tension as a Function of Sodium Chloride

0.1 $\underline{M}$ Sodium Oleate, pH 10.4, vs $\underline{n}$-undecane

Sodium Chloride (m/l)	Interfacial Tension (dynes/cm)
0.01	0.33
0.1	0.26
0.2	0.23

Measurements with a commercial petroleum sulfonate surfactant are
included in Figure 4 for comparison with sodium oleate.

3. Effect of pH. Alkyl carboxylic acids are weak electrolytes, the dissociation constant in general becoming smaller with
increasing molecular weight. It therefore seemed appropriate to
determine the effect of the pH of the aqueous solution on interfacial tension.

Figure 5 summarizes the results for two concentrations of
sodium oleate of varying the pH on interfacial tension. One trend
is that the lower the solution pH the lower the interfacial tension. Harkins et al (16) have demonstrated that the interfacial
tension between benzene and aqueous sodium oleate is lower when
the benzene layer contains oleic acid. Since the partition co-
efficient of oleic acid between alkane and aqueous sodium oleate
greatly favors the alkane phase (17), any oleic acid from hydroly-
sis would tend to partition into the alkane phase. Thus lowering
the solution pH would consequently increase the concentration of
oleic acid in the alkane phase and bring about lower interfacial
tension.

Another trend in Figure 5 is that the 0.01 $\underline{M}$ sodium oleate
is more sensitive to pH effects than the 0.1 $\underline{M}$ sodium oleate.
Mansfield (17) found that an aqueous solution of high ionic
strength or of high pH inhibited the transfer of oleic acid from
paraffin oil to the alkaline aqueous phase. The higher interfacial tensions of the 0.01 $\underline{M}$ sodium oleate solutions in comparison with 0.1 $\underline{M}$ sodium oleate at pH above 10.5 may reflect a
lower concentration and ionic strength. Or if, as Mansfield
suggests, kinetics of transfer between phases at high pH is slow,
equilibrium may not have been reached in these experiments.

In Table III the results of adding excess acid to sodium
oleate solutions are given. These data are not directly comparable to Figure 5 since the aqueous phase contains cosolvent and
the solutions were preequilibrated before testing. However, the
trend to lower interfacial tension with increasing additions of
oleic acid (and thus lower pH) is parallel to Figure 5.

Table III

Interfacial Tension as a Function of Added Oleic Acid

0.01 M Sodium Oleate, 0.1 M Sodium Chloride,

5% Isobutyl Alcohol Pre-equilibrated with

Equal Volume of n-undecane

Oleic Acid Added (m/l aqueous)	Interfacial Tension (dynes/cm)
0	0.65
0.001	0.53
0.002	0.42
0.005	0.48
0.008	0.32
0.01	0.23

4.   Effects of Alcohols. Alcohols are common additives to many surfactant formulations being considered for oil recovery. Wade (12) has studied the effect of alcohol additions on interfacial properties and phase behavior for pure alkyl benzene sulfonates.

Salter (18) divides the alcohols into three main groups on the basis of their oil/water solubility. Isopropyl alcohol is of the water soluble group, isobutyl alcohol is of the intermediate group, and 2-hexanol is of the oil soluble group. Table IV gives some results for alcohols of each class. No strong effect was seen. We have previously reported (15) on the ability of alcohols to reduce the viscosity of sodium oleate gels.

Table IV

Effect of Added Alcohols

Sodium Oleate, 0.1 M Sodium Chloride, pH 10.4

vs. n-undecane

Sodium Oleate (m/l)	Co-Solvent (% aqueous)	Interfacial Tension (dynes/cm)
0.01	None	0.19
	5% Isobutyl alcohol	0.62
0.1	None	0.26
	1% Isopropyl alcohol	0.33
	1% 2-Hexanol	0.17

5.   Effect of Unsaturation. Alkane scans for C_{18} carboxylic acid salts are compared in Figure 6. The alkane position of minimum interfacial tension (n_m) increases with the degree of unsaturation. Increasing interfacial tensions are found in the order oleate, linolenate, linoleate. Sodium stereate, the corresponding

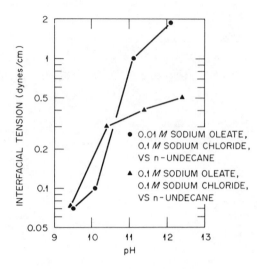

Figure 5.   *Interfacial tension—pH*

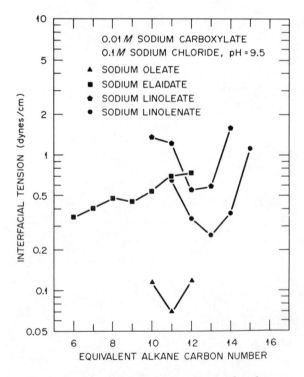

Figure 6.   *Interfacial tension—sodium carboxylates*

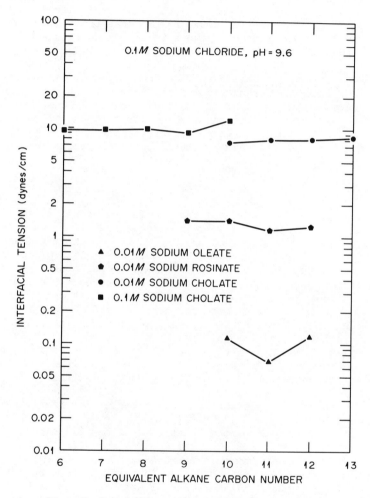

Figure 7.   *Interfacial tensions—cyclic carboxylate acids*

saturated isomer, is too water insoluble to test. Klevens (13) found that increasing unsaturation brought about a small increase in the cmc of fatty acid salts of the same chain length. Comparison of oleate with elaidate gives an idea of the effect of cis-trans isomerism. Sodium eleaidate gave high interfacial tensions over the alkane range tested. Studies on the molecular packing of fatty acids in monolayers indicate that the trans isomer (elaidic acid) forms condensed films with greater ease than the cis isomer (oleic acid((19). Sodium elaidate is also much less water soluble than sodium oleate.

      6. Rosinate and Cholate. These carboxylates were included because of their different (from fatty acid) structure; rosin acids compose about half of the tall oil acids and cholic acid is a representative bile acid that is important in the animal metabolism of fats. Salts of these acids had interfacial tensions that were significantly higher than oleate; no minima were found (Figure 7).

Discussion

      Pure carboxylate salts of fatty acid components of tall oils lower interfacial tensions between aqueous solutions and hydrocarbons. Sodium oleate, used preponderantly in the measurements, exhibits minima in interfacial tension as a function of molecular weight of alkanes as well as other behavior analogous to petroleum sulfonates and pure alkyl benzene sulfonates. Previous investigations (16,20) have shown interfacial tensions in the low millidyne region between some hydrocarbons and aqueous solutions in presence of carboxylates. However, the minimum interfacial tensions attained under conditions we have tried are over an order of magnitude higher than the millidyne values thought necessary for enhanced production of oil. Values obtained with the mixtures of several acids occurring in commercial tall oils were not lower.

      In view of the fact that one would expect interference with surfactant action by Ca(II), Mg(II), and other multivalent ions to be more severe with carboxylates than with sulfonates, the high interfacial tensions observed are discouraging from a practical point of view. However, there are interesting effects of structure between the salts studied, in particular between the cis and trans isomers, elaidate and oleate, and between compounds of different degrees of unsaturation. We are now preparing derivatives of these acids, in order to get more information on the effect of minor structure modifications. In addition, the possibilities of beneficial effects of cosurfactants have been as yet little explored. We believe it premature to conclude that carboxylate surfactants are of no utility.

ACKNOWLEDGEMENT

Research sponsored by the Division of Oil, Gas and Shale
Technology, U. S. Department of Energy under contract W-7405-
eng-26 with the Union Carbide Corporation.

LITERATURE CITED

1. Hoyt, C. H. and Goheen, D. W., "Lignins," Sarkanen, K. V.
    and Ludwig, C. H., eds., Wiley-Interscience, New York,
    N. Y., p. 833 (1971).
2. Anonymous, "Acintol Tall Oil Products," Arizona Chemical Co.,
    30 Rockefeller Plaza, New York, N. Y.
3. McBain, M. E. L. and Hutchinson, E., "Solubilization,"
    Academic Press, New York, 1955.
4. Shinoda, K., Nakagawa, T., Tanamuchi, B., and Isemura, T.,
    "Colloidal Surfactants," Academic Press, New York, 1963.
5. Gillberg, G., Lehtinen, H., and Friberg, S., J. Colloid
    Interfac. Sci. (1970) 33, 40.
6. Shah, D. O., Tamjeedi, A., Falco, J. W., and Walker, Jr.,
    R. D., A.I.C.H.E. J. (1972) 18(6), 1116.
7. Ayers, Jr., R., U. S. Patent 3,616,853 (1971).
8. Purre, H., U. S. Patent 3,362,474 (1968).
9. Williams, S. A., U. S. Patent 3,303,879 (1967).
10. Cayias, J. L., Schecter, R. S., and Wade, W. H., ACS Sym-
    posium Series Number 8, "Adsorption at Interfaces," p. 234,
    1975.
11. Gash, B. and Parrish, D. R., J. Pet. Tech. (1977) 29, 30.
12. Wade, W. H., Morgan, J. C., Schecter, R. S., Jacobson, J. K.,
    and Salager, J.-L., Paper SPE 6844, 52nd Annual Fall Tech-
    nical Conference, SPE-AIME, Denver (1977).
13. Klevens, H. B., J. Amer. Oil Chem. Soc. (1953) 30, 74.
14. Merrill, R. C. and Getty, R. J., J. Phys. and Colloid Chem.
    (1948) 52, 774.
15. Baldwin, W. H., Compere, A. L., Griffith, W. L., Harrison,
    N. H., Johnson, Jr., J. S., Neal, G. W., Smith, D. H.,
    Westmoreland, C. G., Chemicals for Enhanced Oil Recovery,
    Report to Department of Energy/Oil, Gas, and Shale Tech-
    nology, April 23, 1976 - April 22, 1977.
16. Harkins, W. D. and Zollman, H., J. Amer. Chem. Soc. (1926)
    48, 69.
17. Mansfield, W. W., Aus. J. of Sci. Res., Phys. Chem. Sec.
    (1952) Ser. A5, 331.
18. Salter, S. J., Paper, SPE 6843, 52nd Annual Fall Technical
    Conference, SPE-AIME, Denver (1977).
19. Schneider, V. L., Holman, R. T., and Burr, G. O., J. Phys.
    Colloid Chem. (1949) 53, 1016.
20. Seifert, W. K. and Howells, W. G., Anal. Chem. (1969) 41(4),
    554.

RECEIVED September 14, 1978.

Experimental Thermochemistry of Oil Recovery Micellar Systems

R. L. BERG, L. A. NOLL, and W. D. GOOD

U.S. Department of Energy, Bartlesville Energy Technology Center, Bartlesville, OK 74003

The U.S. Department of Energy Bartlesville (Okla.) Energy Technology Center has a comprehensive program of supporting research for enhanced oil recovery. As part of this research, methods are being developed to improve the characterization of surfactant interactions which occur in oil-field flooding. The thermodynamic properties are being investigated under various background conditions of brine, cosurfactant, rock surfaces, and temperatures.

The initial work at Bartlesville has concentrated on measurements of enthalpy changes from dilution and adsorption for surfactant systems. From the observed dilution enthalpy changes, critical micelle concentrations have been determined, and standard state enthalpies of micellization have been calculated. In the studies on adsorption, several properties are of interest: the enthalpy of adsorption, the amount of surfactant adsorbed, the surface area of the solid and determining whether the adsorption is reversible. The kinetics of adsorption and desorption are also of interest.

In the work presented here, these processes have been studied primarily by calorimetry. Planned measurements of partial specific heat and partial molal volume will give additional thermodynamic data on the structure of micellar systems. Heat capacity measurements will allow "simple" extrapolation of measured enthalpy terms to higher temperatures. In addition, a direct measure of the effect of temperature variation is of interest for solution structure studies. Partial molal volume measurements give information on the packing of surfactant monomers and micelles within the water structure. The effect of cosurfactants on the partial molal volume will be of particular interest.

Experimental

Compounds of particular interest for these studies are of the general type used in oilfield flooding. These surfactants are generally complex mixtures of alkyl and alkyl aromatic sulfonates. The compounds discussed in this paper are sodium dodecyl sulfate and the sodium salts of decyl and dodecyl sulfonic acids. Sodium dodecyl sulfate was used as an initial test surfactant for the calorimeter system. The comparison of dodecyl sulfate with dodecyl sulfonate indicated the effect of change of the head group on the properties of the surfactant. The comparison of dodecyl sulfonate with decyl sulfonate showed the effect of varying the alkyl chain length. It was expected that considerable differences would be observed for structural changes of the surfactant. A commercial sample of sodium dodecyl benzene sulfonate was used for the first studies of adsorption phenomena. This material is a mixture of isomers and has been used to obtain values for enthalpies of adsorption on well-characterized surfaces such as silica gel.

Apparatus. The apparatus used in the solution calorimetric study has been previously described in detail (1,2). Briefly, the instrument is a heat-conduction-type flow calorimeter with a power resolution of 0.2 microwatt. Measurements were made at 20°, 25°, 30°, and 35° C ±0.01° C.

Two types of measurements are made with the adsorption calorimeter, also previously described (3). In the batch mode a dry-solid surface is covered with a solution. In the flow mode the enthalpy changes result from a solution flowing through a bed of adsorbent. The flow system uses an LKB 10200 Perpex pump (reference to specific trade names does not imply endorsement by the Department of Energy) with a flow rate of approximately 12 g h^{-1}. Because the silicone tubing on the pump may adsorb surfactant, the pump is placed at the output of the flow system and draws the solution through the cell. An Altex six-way valve is at the input of the flow system, and any one of six solutions can be selected to flow through the cell. Minimum detectable heat pulse is 4.5×10^{-5} cal for the batch and minimum power output is 2.4×10^{-7} cal sec^{-1} for the flow mode. Measurements reported for the adsorption study were made at 25° and 30° C ±0.05° C.

Materials. The studies reported here have been made with monoisomeric sodium dodecyl sulfate, sodium dodecyl sulfonate and sodium decyl sulfonate prepared by Professor E. J. Eisenbraun of Oklahoma State University. The sodium dodecyl benzene sulfonate was a commercial sample purchased from Research Organic/Research Inorganic Chemical Corp. All materials used for background solutions were prepared from reagent grade materials, and the water was distilled from permanganate. Solutions were prepared on a mass basis with corrections for buoyancy.

Two solid surfaces were investigated in the adsorption work, silica gel–Davison No. 62 and solid glass beads. Equilibrium studies have been made on the materials (4), and specific surfaces have been measured.

Strategy. The enthalpy effect measured in the studies of micellar solution structure is due to dilution and demicellization. This can be expressed by

$$\Delta H(\text{meas}) = L_\phi(\text{final concentration}) - L_\phi(\text{initial concentration}) \qquad 1)$$

where the L_ϕ terms are relative apparent molar enthalpies. These terms include contributions from dilution, demicellization, and ion pairing.

In adsorption studies, the flow system measures the enthalpy of adsorption of a material whose surface is wet. The measured enthalpy includes, among other terms, the enthalpy of replacement of the water by the adsorbate. The adsorbent may be exposed to a series of solutions of increasing or decreasing concentration, thus approximating the differential enthalpy of adsorption; or it may be exposed to a sharp increase in concentration to determine integral enthalpies of adsorption. The batch system is designed for determining enthalpies of immersion of a solid into solutions. For all the calorimetric studies, the measured enthalpy is determined by

$$\Delta H(\text{meas}) = \mathcal{E} \times S/n \qquad 2)$$

where $\mathcal{E}$ is a calibration constant, S the measured signal, and n the amount of material reacting.

Results and Discussion

Solution Work. Results of measurements of enthalpies of the surfactants are shown in Figures 1 through 7. The observed critical micelle concentrations are tabulated in Table I. For several surfactant–cosurfactant systems, the surfactant was not sufficiently soluble to allow determination of the critical micelle concentration.

The lack of data reported in the literature prevents comparing thermodynamic measurements of micellization enthalpies. Although numerous calorimetric studies have been made, many times the measured enthalpies have been mathematically manipulated to give a value for the enthalpy of micellization. Although valuable in testing theories, such manipulation obscures the use of different methods for treating data. As a result, each method may have produced a different value. The methods briefly summarized are as follows:

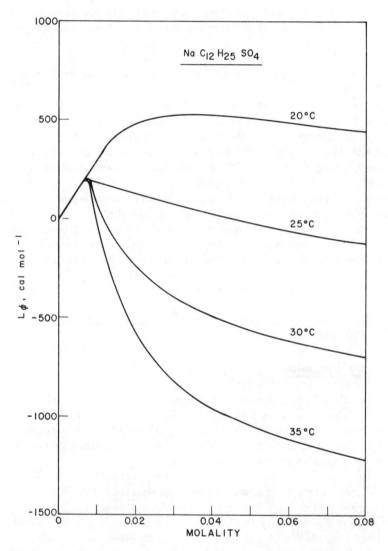

Figure 1.   Sodium dodecyl sulfate enthalpy curves at various temperatures

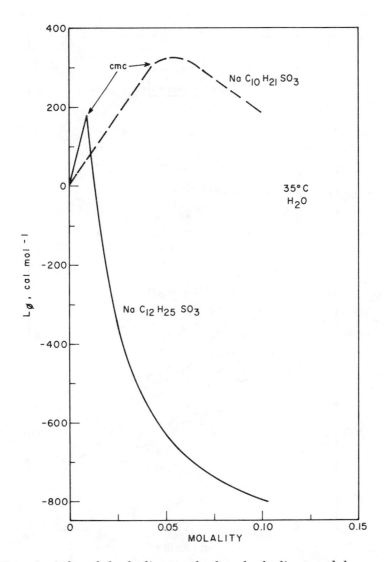

Figure 2. Sodium dodecyl sulfonate and sodium decyl sulfonate enthalpy curves at 35°C

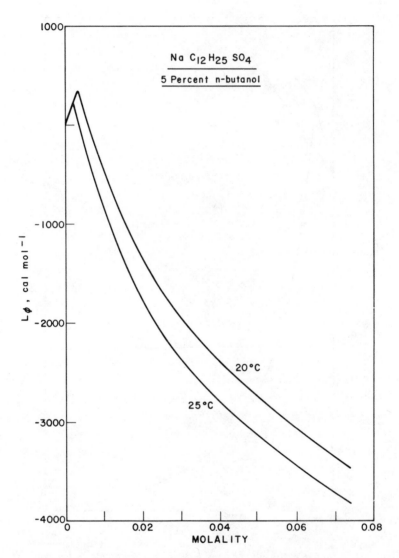

Figure 3.   Sodium dodecyl sulfate enthalpy curves in 5% n-butanol

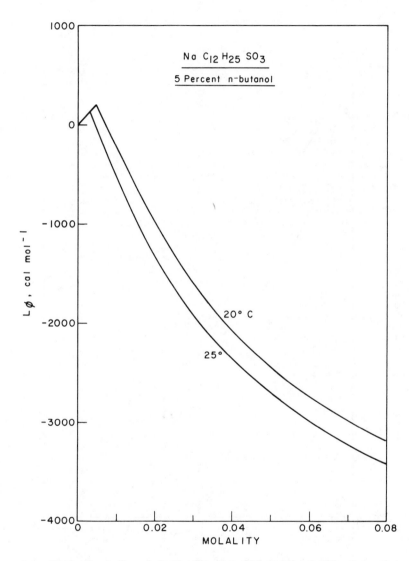

Figure 4.   Sodium dodecyl sulfonate enthalpy curves in 5% n-butanol

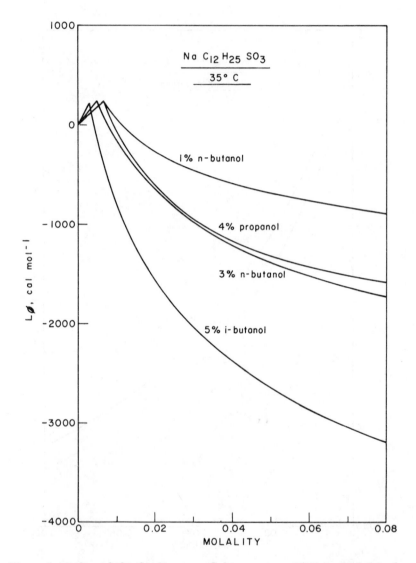

Figure 5. *Sodium dodecyl sulfonate enthalpy curves at 35°C in alcohol backgrounds*

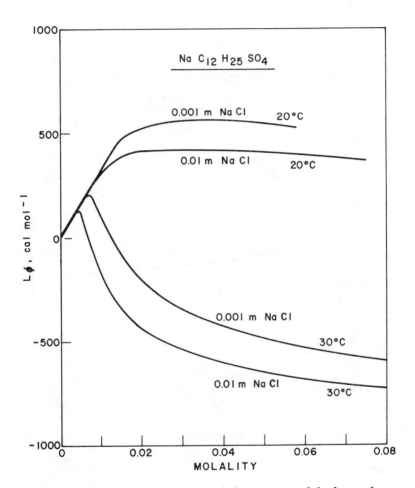

Figure 6. Sodium dodecyl sulfate enthalpy curves in salt background

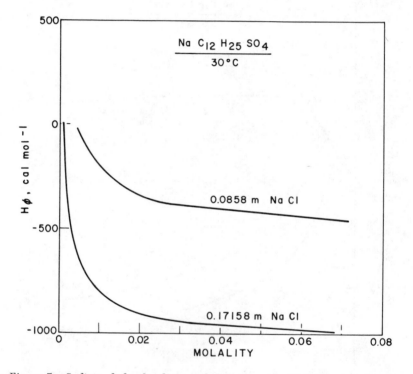

Figure 7.   Sodium dodecyl sulfate enthalpy curves in high-salt concentration background

TABLE I

CRITICAL MICELLE CONCENTRATIONS

$$\text{cmc, mol(kg } H_2O)^{-1} \times 10^3, \pm 0.5$$

Sodium dodecyl sulfate

Background	20° C	25° C	30° C	35° C
5% n–Butanol	3.6	2.5	–	–
H_2O	8.1	8.3	8.4	8.3

Sodium dodecyl sulfonate

Background	20° C	25° C	30° C	35° C
5% n–Butanol	4.9	3.2	–	–
5% i–Butanol	7.2	6.0	3.8	3.0
3% n–Butanol	–	–	7.5	5.2
1% n–Butanol	–	5.5	6.2	6.3
4% n–Propanol	–	–	8.9	6.5
H_2O	–	–	–	10.2

Sodium decyl sulfonate

Background	20° C	25° C	30° C	35° C
5% n–Butanol	9.5	–	–	12.8
5% i–Butanol	9.8	–	–	16.6
3% n–Butanol	13.5	–	–	13.2
1% n–Butanol	–	–	–	38.0
4% n–Propanol	23.8	–	–	30.0
H_2O	–	–	–	41.8

$$L_\phi(\text{above cmc}) - L_\phi(\text{below cmc}) , \qquad\qquad 3)$$

$$L_\phi(\text{above cmc}) - L_\phi{}^\circ , \qquad\qquad 4)$$

$$L_\phi(\text{above cmc}) - \Delta H(\text{ion-pairing}) - L_\phi(\text{dilution}) , \qquad 5)$$

$$\overline{L}(\text{above cmc}) - \overline{L}(\text{below cmc}) , \qquad\qquad 6)$$

$$\overline{L}(\text{above cmc}) - \overline{L}^\circ , \qquad\qquad 7)$$

$$\overline{L}(\text{above cmc}) - \Delta H(\text{ion-pairing}) , \qquad\qquad 8)$$

where $\quad \overline{L} = L_\phi + \dfrac{1}{2} m^{\frac{1}{2}} (dL_\phi / dm^{\frac{1}{2}}) . \qquad\qquad 9)$

The $\overline{L}$ terms are partial molal enthalpies defined by Equation 9. Another method which calculates the enthalpy change for adding 1 mole of monomer to the micelle has also been used. Additional explanation of these methods can be found in the literature (1,5).

Table II lists the results from our study in which we used relationship 6. For each study in which water was used, the $\overline{L}$ term reached a nearly constant value soon after the critical micelle concentration. In alcohol background such a constancy was not observed. In salt backgrounds (Figures 6 and 7) the leveling occurred even more rapidly.

TABLE II

TENTATIVE VALUES FOR THE ENTHALPY OF MICELLIZATION
IN H_2O AT 35° C FROM EQUATION 6

	ΔH, cal mol^{-1}
Sodium dodecyl sulfonate	−1440
Sodium decyl sulfonate	−140
Sodium dodecyl sulfate	−1780

Only a slight temperature dependence was noted for the cmc. The temperature dependence of the enthalpy terms was large. This indicates that calorimetry is a sensitive tool for detection of changes in solution structure, and implies that temperature control is critical for studying the properties of micellar systems.

The premicellar region for the three surfactants studied showed a steeper slope than the expected 1:1 electrolyte slope. A possible explanation of this slope by the existence of a dimer in solution has been previously discussed (1). Additional work in this area is required.

The comparison studies of decyl and dodecyl sulfonates show the strong dependence of the enthalpy terms on chain length. The cmc for the decyl salt is considerably higher under all conditions than the dodecyl salt. The planned heat capacity and partial molal volume measurements will be of interest in this comparison.

Adsorption Work. In the adsorption work, glass beads were used as a blank to determine whether the heat of dilution of the surfactant or changes of flow pattern lead to a thermal effect. At both 25° and 30° C, the heat effect of flowing an increasing concentration from 0.1 to 2% in five steps was not detectable. Since the amount of adsorption to solid glass beads is small, it can be concluded that there is either very little mixing of the two solutions as the interface between them moves through the cell, or if there is mixing that the heat effect is too small to affect the adsorption measurements.

Table III gives typical results from an adsorption run. In all cases the heat effects were small, which suggests that only certain sites bind the surfactant and not the entire surface. In addition, there was no evidence that the sites on the surface were saturated, even when a 2% surfactant solution was used.

TABLE III

ENTHALPY OF ADSORPTION OF SODIUM DODECYL BENZENE SULFONATE ON SILICA GEL

$-\Delta H$(adsorption), cal $\times 10^3$ (g Davison 62)$^{-1}$

	25° C	30° C
$H_2O \rightarrow 0.1\%$	5.5	5.5
$0.1\% \rightarrow 0.2\%$	4.8	4.5
$0.2\% \rightarrow 0.4\%$	5.7	4.8
$0.4\% \rightarrow 1.0\%$	11.0	8.4
$1.0\% \rightarrow 2.0\%$	13.4	9.3
$2.0\% \rightarrow H_2O$	-41.6	-38.0

Within the experimental limits there is little temperature dependence for the lower concentrations. However, no firm conclusions can be drawn without running the system at other temperatures. An interesting comparison from Table III is the following:

$-\Sigma \Delta H$, cal $\times 10^2$ g^{-1}

	25° C	30° C
Adsorption	40.4	32.5
Desorption	-41.6	-38.0

The results show fairly good additivity for 25° C and poor additivity at 30° C. The pattern, also shown in Table IV, may be explained by considering the shape of the observed enthalpy curves, Figure 8. The results of allowing the surfactant to flow through glass beads show that the above results need no corrections for mixing of the solutions, and suggest that when the concentration changes, one solution replaces the other completely. The shape and width of the curve show that adsorption onto the surface is slow with respect to the flow of liquid through the cell. About 1.5 minutes is

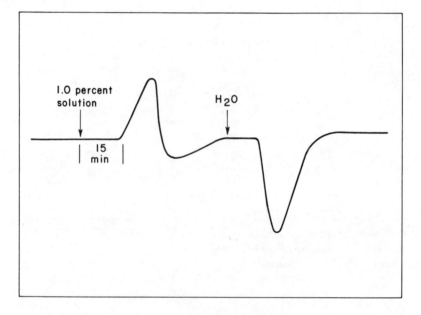

Figure 8.   Flow adsorption calorimeter enthalpy curve

TABLE IV

ENTHALPY OF ADSORPTION AND DESORPTION OF SODIUM
DODECYL BENZENE SULFONATE ON SILICA GEL AT 30° C

$-\Delta H$, cal $\times 10^3$ (g Davison 62)$^{-1}$

	Adsorption	Desorption
0.1%	5.5	-6.2
0.2%	10.0	-11.2
0.4%	14.8	-16.7
1.0%	23.2	-30.4
2.0%	32.5	-38.0

required to allow sufficient liquid to flow to fill the cell, whereas the curve
shown represents 35 minutes. The peak followed by a long tail suggests that
not all the sites are equivalent either kinetically or energetically--in
simple terms the surfactant is adsorbing from a solution containing micelles.
The net reaction is taking surfactant from micelles and binding them to the
surface. The heat of replacing the water molecules from the surface is
diminished by the heat of demicellization. This is clearly seen in the dip
at the end of the pulse, where the net heat effect is endothermal; the heat
of demicellization is greater in magnitude than the heat of adsorption.
When water is returned to the cell it flushes micelles from the cell faster
than desorption can occur, and the heat effect is taking the surfactant from
the surface to a solution concentration below the cmc.

This procedure is not a good test of the reversibility of binding, since
the reaction is not a simple reversal of the process within the space of the
calorimeter. We are now experimentally checking the reversibility of
binding by alternating flows of solutions of comparable concentration--that
is, 2% solution, then 1% solution, then 2% solution--so that the concentration never falls below the cmc. It may be possible to deduce some
kinetic information from the shape of the curve, but measurements of the
concentration of surfactant near the outlet of the cell would give this
information more readily. Considerable additional work is needed in this
area.

Summary

This paper summarizes the on-going research program in thermochem-
istry of oil recovery at the Bartlesville Energy Technology Center. Information has been gained on the structure of micellar solutions and of interactions
with surfaces. In order for thermodynamics to make a greater contribution

to the understanding of oil recovery considerable additional studies at
Bartlesville and in other laboratories must be made with idealized systems
and with real systems. This report is only a sketch of what could be investigated. The understanding of the mechanisms of surfactant loss by
adsorption, micellar system stability, and phase behavior will be of great
value in the improvement of flooding processes for enhanced recovery of
petroleum.

Literature Cited

1   Berg, Robert L., ERDA BERC/TPR-77/3, August 1977, 35 pp.

2   Berg, R. L. and Good, W. D., Proc. 2nd ERDA Symposium on
     Enhanced Oil and Gas Recovery, Tulsa, Okla., Sept. 9-10, 1976,
     Vol. 1, Petroleum Publishing Co., Tulsa, Okla.

3   Berg, R. L., Noll, L. A., and Good, W. D., Proc. 3rd ERDA
     Symposium on Enhanced Oil, Gas Recovery, and Improved Drilling
     Methods, Tulsa, Okla., Aug. 30-31, Sept. 1, 1977, Vol. 1,
     Petroleum Publishing Co., Tulsa, Okla.

4   Walker, R. D., University of Florida, Improved Oil Recovery
     Semi-Annual Report, December 1976.

5   Desnoyers, J. E., De Lisi, R., Ostiguy, D., and Perron, G.,
     to be published in "Solution Chemistry of Surfactants" edited by
     K. L. Mittal and A. S. Kertes, Plenum Press.

RECEIVED August 16, 1978.

Participation of Selective Native Petroleum Fractions in Lowering Interfacial Tensions of Aqueous Alkaline Systems

P. A. FARMANIAN
AMINOIL USA, Huntington Beach, CA 92645

N. DAVIS, J. T. KWAN, R. M. WEINBRANDT, and T. F. YEN[1]
University of Southern California, Los Angeles, CA 90007

The presence of carboxylic acids (1), phenols (2), or porphyrins (3) in petroleum is believed to be beneficial to oil recovery in the presence of dilute base due to the low interfacial tension exhibited. It has been recognized that the first type of compound is responsible for the recovery efficiency of alkaline water flooding (4). Seifert and Howells (5) fractionated the many acid components of the California crude oil into minute quantities to ascertain their interfacial activities. However, none of the isolated fractions revealed surface tensions lower than the original crude oil. Furthermore, wherever the surface tension is significantly lower for a few fractions, the value is highly dependent on the alkaline concentrations. Hence, the exceedingly narrow range of pH levels precludes any given practical use.

Apparently, higher acid number of a given crude, either by nature or due to the addition of known acid, would lower its interfacial tension. However, Cooke, Williams, and Keledzne (4) found that although in-situ oxidation with air further increases the acid number of a given crude, this artificially-made high acid number crude could not be successfully flooded with alkaline water.

The caustic method as a means of improved waterflooding for enhanced oil recovery is a complex process. Johnson has outlined four recovery mechanisms (6). Presumably, besides the ultralow tension mode, there are other requirements to ensure efficient and stable recovery of an oil in a given reservoir. To name a few: spontaneous emulsification, entrainment, entrappment, wettability reversal in both directions, etc. In order to maintain a particular set of pro-

---

[1]To whom all correspondence should be addressed, USC, Dept. of Chemical Engineering.

perties, the component which is responsible for the particular
activity in crude petroleum should be examined. For example
it is possible to have a particular fraction in crude as a co-
surfactant, or as a negative surfactant (counter surfactant) besides the active surfactant. It is also possible to identify
a particular fraction of crude as responsible for maintaining
a stable emulsion or regulating the flow of the emulsion. A
possible component is the resin or asphaltene cut.

Two approaches can be used to study the nature of a mixture consisting of a multi-component collection of various
molecules. The first is to select a sample model compound
which possesses certain physical properties for simulating
the behavior of the mixture. The second approach is to separate the complex mixture into gross fractions and to inves-
tigate the properties of each individual fraction for synthe-
sizing the properties of the composite. The first approach
has been used by many investigators. Cash and co-workers (7)
used the concept of equivalent alkane carbon number (EACN) for
modeling crude oils. Radke and Somerton (8) used synthetic
systems (oleic acid in Ottawa Sands packs) for caustic flooding
research. To the knowledge of the authors, little work has
been done and reported on the second approach.

Hence, we conducted the present preliminary work of
separating production crude into gross fractions at lower
temperatures. Each fraction is separately examined for
interfacial tension and associated properties.

Experimental

Sample description

For our experiments, we utilized fresh crude petroleum
samples from AMINOIL USA's production wells at the Huntington
Beach field (figure 1). The Huntington Beach field is a
major oil accumulation lying on the California coastline
approximately 20 miles southeast of Los Angeles. The field
has a length of seven miles along Newport-Inglewood Fault Zone
and a maximum width of three miles. Production in the off-
shore area is from five major zones with the upper zone
assisted by steam injection while three of the lower zones are
under waterflood and the remaining lower zone is producing on
primary. The samples that we employed were produced from the
lower Main Zone reservior (Well No. S-47).

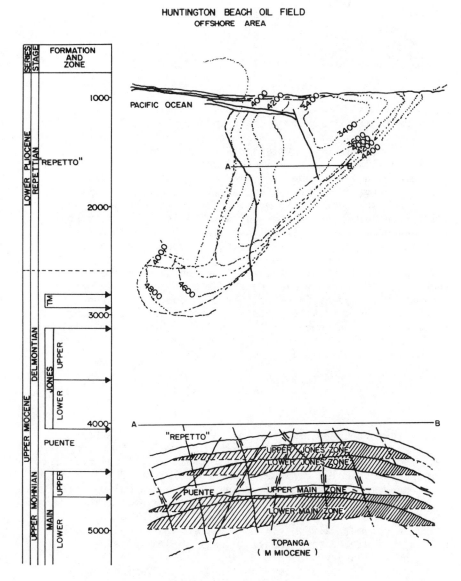

Figure 1. *Various production zones at the Huntington Beach field*

Region	Well No.	Recovery Stage	Depth. (ft.)
Upper Zones	UJ-255	Water Flood	3500
Upper Main Zone	426-104	Primary	4500
Lower Main Zone	S-47	Water Flood	4800

Recovery from this reservoir is by water flood. This petroleum
is representative of a typical California crude. The crude was
initially distilled at atmospheric pressure to remove excess
water and low-boiling volatile components (Distillate cut #1).
The remainder was vacuum distilled at 0.4 mm Hg. to yield
Distillate cut #2 (46 to 110°C) and Distillate cut #3 (118 to
156°C). Distillate cut #3 was further fractionated by utilizing open column chromatography with silica gel. This separation scheme gave two fractions: (a) a benzene-eluted fraction and (b) an ether-eluted fraction. The benzene-eluted
fraction was composed of 90% by weight of the original Distillate cut #3 and the ether-eluted fraction was composed of the
remaining 10%. The remaining residue after distillation was
subjected to solvent fractionation into gas oil (pentane-
soluble, propane-soluble), resin (pentane-soluble, propane-
insoluble), asphaltene (pentane-insoluble, benzene-soluble)
carbene plus carboid (benzene-insoluble) according to the
scheme for coal liquid (9) and shale oil (10).

Properties

The measurement of interfacial tension (IFT) was performed
through the spinning drop method as described by Cayias (11).
All measurements were conducted in an aqueous alkaline solution
of sodium orthosilicate unless otherwise indicated. The aqueous
for all samples also had a 7500 ppm NaCl concentration, this
being an optimum salt concentration to minimize sodium orthosilicate requirements. In all cases, the drop size was consistent and the data repeatable.

Proton NMR spectra were obtained from a Varian T-60 spectrometer. The solvent used was 99.8% $CDCl_3$ containing trime-
thylsilane as a marker. Infrared spectra were obtained with
a Beckman Acculab-6 spectrometer. The elementary analyses
were determined by Elek Microanalytical Laboratories, Torrance,
California. A Waters high pressure liquid chromatograph was

also used to study the changes in composition.

Results

The weight recovery of all bulk fractions is listed in Table I. Evidently, the ratio of the volatiles to non-volatiles of the two samples which have undergone water flooding is consistent (3:1), when compared with that (2:1) of the primary recovery. In order to test the consistency of all the fractions, both NMR and IR techniques were applied. In Figure 2, the H/C atomic ratio obtained from elemental alnalysis is plotted vs. the carbon aromaticity, f_a, obtained from NMR (12,13) and the extent of hydrogen saturation (h_s) obtained from IR (14). The linearity within the fractions suggests that no cracking or condensation occurred during the separation process.

TABLE I

PERCENT OF BULK FRACTIONATION OF PETROLEUM SAMPLES

	Primary	Waterflood	
	426-104	S-47	UJ-255
Distillate I	8.1	16.8	16.8
Distillate II	7.1	25.1	10.1
Distillate III	50.4	30.5	48.2
Non-volatile			
Gas	11.3	12.6	6.9
Resin	14.3	0.7	---
Asphaltene	3.6	7.7	2.7
Benzene-insoluble	----	0.2	9.6

Surface tension data were measured in both distilled water and 7500 ppm salt solution with an alkalinity of sodium ortho-silicate ranging from 60 to 60,000 ppm. As an example, fractions of samples S-47 of the Lower Main Zone were investigated.

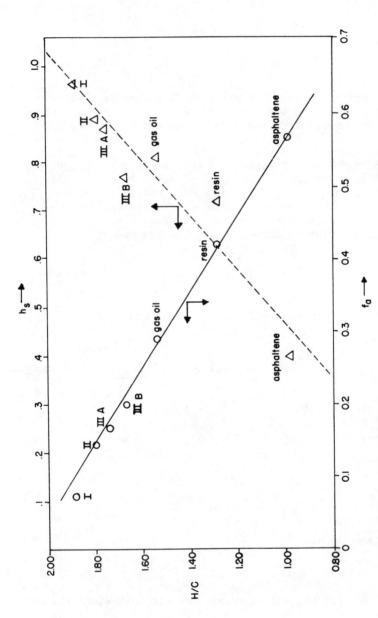

Figure 2. Chemical properties of all the fractions isolated. The carbon aromaticity (f_a) is obtained from NMR. The extent of saturated hydrogens (h_s, $h_s = 1 - h_a$) is obtained from IR.

Figure 3 compares the interfacial tension of Distillate cut #2 and Distillate cut #3 to the original Lower Main Zone crude. Distillate cut #1 showed no IFT activity. The interfacial tension measurements in Figure 3 are the minima of a time dependent function. It is important to note that the low-tension values of the original crude are exceedingly narrow in pH range. However, for Distillate cut #3, this narrow gate has been widened considerably. It is to be noted that its interfacial tension is lower than that of the original crude.

Figure 4 compares the Distillate cut #3 chromatographed fractions to the original Distillate cut #3. The ether-eluted fractions had a density greater than water and therefore we were unable to obtain IFT data using the spinning drop method. However, the IFT activity of a reconstructed cut containing 95% of the benzene-eluted fraction and 5% of the ether-eluted fraction was examined. The figure indicates that the IFT activity of the benzene eluted cut was insignificant. However, the excellent IFT activity of the original Distillate cut #3 was partially restored by the reconstructed cut. This figure tentatively indicates that the ether-eluted fraction is primarily responsible for the IFT activity of Distillate cut #3.

The time dependence of surface tension at the minimum value has been investigated within the range of 2-40 minutes. The results are summarized in Figure 5.

Discussion

It is obvious from Figure 3 that the ultra-low tension values of the total crude are exceedingly narrow in pH range. For Distillate cut #3, which accounts for 30% of the total crude, the narrow "gate" nature (15) has been broadened considerably in alkalinity range. It is possible that under certain conditions, the very low interfacial tension levels obtainted from the caustic solution-crude oil interaction enabled the oil to be entrained in a continuous flowing alkaline-water phase, resulting in a substantial reduction in residual oil saturation (16,17).

The study of the age of the interface, as illustrated in Figure 4, clearly demonstrates that Distillate #3 undergoes spontaneous emulsification. England and Berg (18) attributed this to the transfer of surfactant across the interface with a large desorption barrier when tension dropped continuously. In the present case, the rapid increase in tension accompanied by the single drop instability may also suggest a rapid coalescence process for emulsification. As pointed out by Schechter and Wade (19), the systems which emulsify and rapidly coalesce

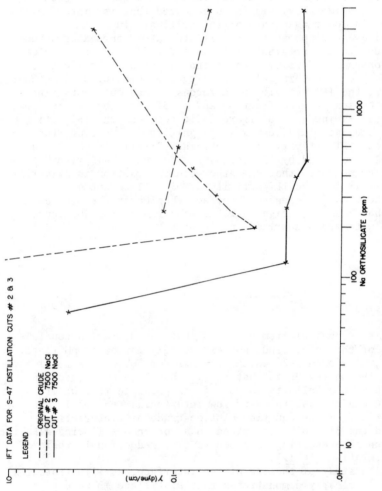

Figure 3. Interfacial tension properties of distillate Cuts #2 and #3 from lower main zone by spinning-drop method

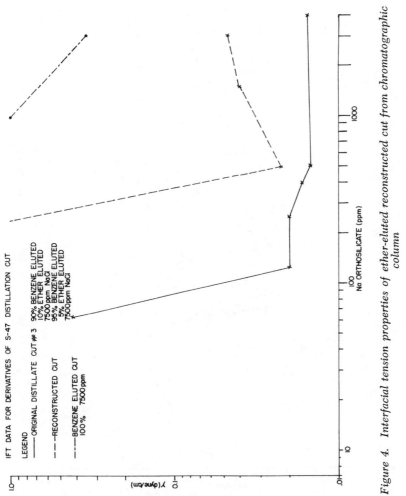

Figure 4. Interfacial tension properties of ether-eluted reconstructed cut from chromatographic column

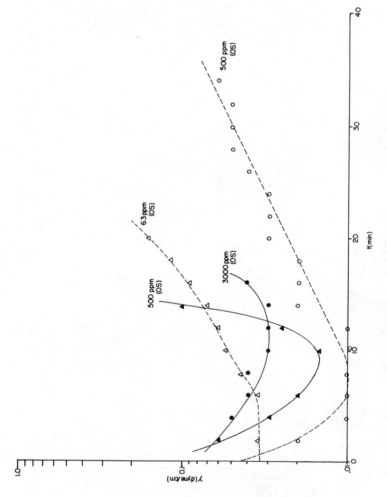

Figure 5.   Age of interphase at the lowest interfacial tension distillate Cut #3

sponstaneously give better oil recovery than those which form
stable emulsions spontaneously.

Distillate cut #3 obtained from bulk separation is still
dark colored which may indicate the presence of small quantities
of asphaltic material (20). Figure 3 indicates that it is
possible that a composite of compounds, besides carboxylic acids,
may be required to yield optimal recovery. In order to understand the mechanism of oil recovery, the contribution of a given
individual fraction to ultra-low surface tension characteristics
and the contribution of various combinations of individual fractions contribute greatly to viscosity behavior. Therefore,
interfacial rheology may be dependent on the appropriate composition of the crude oil.

It is hoped that the preliminary phase of this work will
stimulate other necessary studies so that the correct and
essential mechanisms of recovery can be understood. At this
stage, there are no reliable criteria to decide which crude may
be a good candidate for alkaline water flooding. Hence, iso-
lating components that play a role in alkaline flooding may
provide insight into the mechanism through which alkaline
flooding recovers oil. Identifying essential components in
crude oil is practical for alkaline flooding. In other words,
it is possible to develop a screening device in which the
presence of certain native components in the crude will indicate
the potential for alkaline flooding. Finally it is hoped that
this preliminary work may assist in developing a new class of
naturally-occurring surfactants.

Acknowledgement

A portion of this work is supported by DOE contact
EW-78-S-19-0005 and the NSF Summer High School Student Program
(for ND). We also wish to thank Miss Linda Wong for her
clerical help.

References

1. Seifert, W.K., "Carboxylic acids in Petroleum Sediments,"
   Prog. Chem. Nat. Products, 1975, Springer-Verlag, pp. 1-49.

2. Neumann, H.J., Erdol Kohle, 17, 346 (1964)

3. Dunning, H.N., Moore, J.W., and Denekas, M.O., Inds. Eng.
   Chem. 45, 1759 (1953)

4. Cooke, C.E. Jr., Williams, R.E., and Kolodzie, P.A., J.
   Petrol. Tech. 26(12), 1365 (1974).

5. Seifert, W.K. and Howells, W.G., Anal. Chem., 41(4),
   554 (1969)

6. Johenson, E. C. Jr., J. Petrol. Tech., January 85 (1976)

7. Cash, R.C., Cayias, J.C., Fournier, R.G., Jacobson, J.K., Schares, T., Schechter, R.S., and Wade, W.H., "Modeling Crude Oils for Low Interfacial Tension," SPE 5813, 1976.

8. Radke, C.J. and Somerton, W.H., "Enhanced Recovery with Mobility and Reactive Tension Agents," Proc., Tulsa ERDA Meeting, 1977.

9. Schwager, I., and Yen, T.F., "Separation of Coal Liquids from Major Liquefaction Processes in Meaningful Fractions," in Liquid Fuel from Coal, (R.T. Ellington, ed.), Academic Press, 1977, pp. 233-243

10. Yen, T.F., Wen, C.S., Kwan, J.T., and Chow, E., "The Nature of Shale Oil-Derived Asphaltene" in Oil Sand and Oil Shale (O.P. Strausz, ed.) Springer Verlag, 1978, pp. 309-324

11. Cayias, J.L., Schechter, R.S., and Wade, W.H., "The Measurement of Low Interfacial Tension via the Spinning Drop Technique".

12. Brown, J.K., and Ladner, W.R., Fuel 39, 87 (1960)

13. Yen, T.F., and Erdman, J.G., ACS Preprint, Div. Petrol. Chem. 7(3), 99 (1962).

14. Yen, T.F., and Erdman, J.G., ACS Preprint, Div. Petrol. Chem. 7(1), 5 (1962).

15. Jennings, H.Y. Jr., Johnson, C.E. Jr., and McAuliffe, C.D., J. Petrol. Tech., December, 1344 (1974).

16. Carmichael, J.D., and Alpay, O.A., "Caustic Waterflooding Demonstration Project Ranger Zone Long Beach Unit, Wilmington Field California". Tulsa ERDA Meeting, 1977.

17. Ehrlic, R. and Wygal, R.J. Jr., "Interrelation of Crude Oil and Rock Properties with the Recovery by Caustic Waterflooding", SPE 5380, Tulsa, 1976

18. England, D.C. and Berg, J.C., AIChE J. 17, 313 (1971).

19. Schechter, R.S. and Wade, W.H., "Spontaneous Emulsification and Oil Recovery", ERDA Report EY-76-S-0031, 1976

20. Yen, T.F., Erdman, J.G., and Saraceno, A.J., Anal. Chem. 34, 694 (1962).

Received October 23, 1978.

Spontaneous Emulsification and the Effect of Interfacial Fluid Properties on Coalescence and Emulsion Stability in Caustic Flooding

D. T. WASAN, S. M. SHAH, M. CHAN, K. SAMPATH, and R. SHAH

Department of Chemical Engineering, Illinois Institute of Technology, Chicago, IL 60616

Surfactant or micellar flooding is one of the more promising tertiary oil recovery techniques currently being developed. This process employs an aqueous surfactant slug followed by a mobility buffer (usually polymer-thickened water) to displace oil locked in small pores. The role of the surfactant is to reduce the interfacial tension between the entrapped oil and the flooding fluids. The present state-of-the-art reveals that surfactant selection for a tertiary oil recovery process is made on the basis of ultralow interfacial tension between the crude oil and the aqueous phase. Although such ultralow values of interfacial tension may be necessary to ensure displacement of the oil from the porous rock, this low interfacial tension can, on the other hand, also result in considerable emulsification or dispersion of the oil in the water. The resulting emulsions can be quite stable and therefore, both difficult and costly to separate. A number of recent studies suggested that the poor efficiency of oil recovery was due to emulsion stability problems (1, 2, 3, 4, 5).

We have recently reported (6, 7) that those surfactant formulations which yield good oil recovery exhibit both low interfacial tensions and low interfacial viscosities. Our experiments have shown that surfactant formulations which ensure low interfacial viscosity will promote the coalescence of oil droplets and thereby decrease the emulsion stability, thus enhancing the formation of a continuous oil bank. It has been demonstrated that the requirements for emulsion stability are the presence of an interfacial film of high viscosity and a film of considerable thickness. We have observed that the surfactant concentration which minimizes the interfacial tension may not simultaneously minimize the interfacial viscosity. Hence, our results indicate both interfacial tension and interfacial rheology must be considered in selecting surfactant formulations for tertiary oil recovery.

The idea of utilizing the natural organic acids present in crude oil to produce in-situ surfactants and displace oil by the injection of caustic solutions has been thought to be economically

0-8412-0477-2/79/47-091-115$06.50/0

attractive. There are several proposed mechanisms including
emulsion formation and stability by which alkaline water flooding
may improve oil recovery (8, 9, 10, 11, 12, 13, 14, 15).
This paper presents observations on the difference in behavior of emulsification processes which can occur during surfactant and caustic flooding in enhanced recovery of petroleum.
Cinephotomicrographic observations on emulsion characteristics
generated at the California crude oil-alkaline solution interface
as well as in the Illinois crude oil-petroleum sulfonate system
are reported. The interdroplet coalescence behavior of oil-water
emulsion systems appear to be quite different in enhanced oil recovery processes employing various alkaline agents as opposed to
surfactant/polymer systems.

In this paper we report first the spontaneous emulsification
mechanisms in the petroleum sulfonate and caustic systems. This
is followed by the kinetics of coalescence in alkaline systems
for both the Thums Long Beach (heavy) crude oil and the
Huntington Beach (less viscous) crude oil. Measurements of interfacial viscosity, interfacial tension, interfacial charge and
micellar aggregate distributions are presented. Interrelation-
ships between these properties and coalescence rates have been
established.

Spontaneous Emulsification Mechanisms in Sulfonate and Caustic Systems

The mechanisms of emulsification in oil reservoirs when
chemical slugs are injected into the reservoir are largely un-
known. This study was conducted to understand and characterize
these mechanisms by microscopic observations and high speed cine-
photomicrography.

The spontaneous emulsification mechanisms were determined
for the 3% (Active) Petrostep 420 and 1.5% NaCl and 0.58% n-
hexanol versus non-pre-equilibrated Salem crude oil (Sulfonate
System). The mechanisms were also determined for Thums Long
Beach Crude Oil (Well 108 B) versus 0.05 M (0.2% by weight)
NaOH and 1% NaCl (Caustic System). The process of emulsification
was observed through the Nikon LKe Interference Phase Microscope
(magnifications of 400 and 1,000). The action was captured by
high speed cine-photomicrography (64 frames/second) for the
caustic system and normal speed (24 frames/second) for the sulfonate system.

The procedure used for studying spontaneous emulsification
was the following. A droplet of crude oil was placed on a clean
glass slide. The droplet was covered with a coverslip (0.17 mm
thick). The size of the oil droplet was chosen such that
approximately half of the volume between the cover slip and the
glass slide was occupied by the crude oil. The microscope was
focused at the oil-air interface. A sample of the non-pre-
equilibrated aqueous phase was taken in a pasteur pipet. The tip

of the pipet was placed at the edge of the coverslip and a small
quantity of the aqueous phase was allowed to flood the space between the coverslip and the glass plate which was not occupied by
the crude oil. The aqueous phase thereafter came in contact with
the crude oil under capillary pressure.

Upon contacting of Salem Crude with the sulfonate system a
mutual dissolution of the phases (oil, surfactant, co-surfactant
and water) takes place. The phase which contains the mutually
dissolved (or dispersed) components of the system does not possess
the same phase contrast between it and the aqueous phase as between the crude oil and the aqueous phase as seen from Figure 1A.

The second stage of emulsification as shown in Figure 1B in-
volves the development of a very thick interfacial film around the
mutually dissolved phase. This film is approximately five microns
in thickness (Figure 1C).

The third stage occurs when small droplets of oil start
appearing in the mutually dissolved phase and in the interfacial
film (See Figure 1D). There is coalescence between these small
droplets. The mutually dissolved phase thereafter becomes indis-
tinguishable from the surfactant film.

In the last stage of spontaneous emulsification the interfacial film disintegrates as shown in Figure 1E. In this, the
longest of the stages, the inter-droplet film disintegrates into
surfactant micelles (with or without co-surfactant) and small oil
droplets (which either coalesce with larger oil droplets or are
released into the continuous aqueous phase). This mechanism is
most closely related to the mechanism of diffusion and stranding
(16).

In the caustic system the mechanisms of spontaneous emulsification lead to the formation of both an oil-in-water and a water-
in-oil emulsion. A representative photomicrograph has been included in Figure 2.

The mechanism of formation of the water-in-oil emulsion proceeds in the following manner. First, fingering of the aqueous
phase into the oil phase leads to the formation of large aqueous
droplets inside the oil phase. These large droplets thereafter
exhibit a revolving motion during which they disrupt into small,
more stable aqueous droplets (See Figures 2A, 2B and 2C). Via
spontaneous emulsification by interfacial turbulence threads of
oil are thrown into the aqueous phase where they disintegrate into
droplets as illustrated in Figure 2D.

Another mechanism of formation of the oil-in-water emulsion
was observed. This involved the development of buds of oil at the
oil-aqueous interface which were immediately pinched off to form
an oil-in-water emulsion (See Figure 2E).

Therefore, the mechanism of spontaneous emulsification in
the caustic system is interfacial turbulence. The mechanisms for
the sulfonate system are diffusion and stranding.

Two features of these emulsification systems worth noting
are that the sulfonate system only formed an oil-in-water emulsion

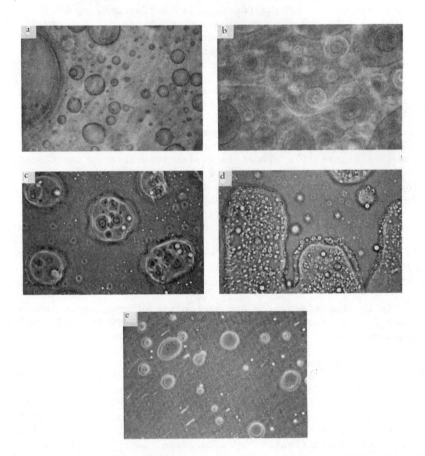

Figure 1.   Emulsification phenomena in Salem crude oil–Petrostep 420 containing n-hexanol: (a) solubilization and diffusion; (b) development of interfacial film; (c) presence of thick films around droplets; (d) coalescence of small droplets; and (e) disintegration of interfacial film.

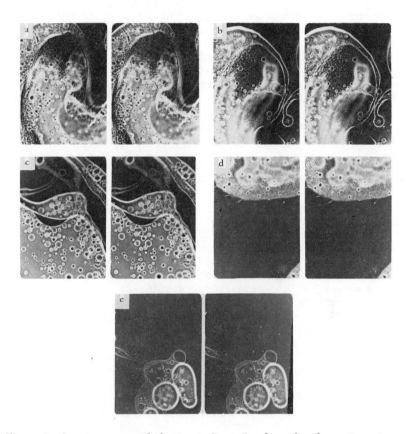

Figure 2. Spontaneous emulsification in Long Beach crude oil–caustic system: (a) initiation of fingering action and formation of water in oil droplets; (b) termination of fingering action and formation of water in oil droplets; (c) threads of oil droplets; (d) formation of very thin strings and appearance of oil droplets in aqueous phase; and (e) appearance of buds of oil at the oil–aqueous interface

while the caustic system formed both oil-in-water and water-in-oil emulsions.
      For the sulfonate system the rate of emulsification was relatively slow and yielded an unstable emulsion. The emulsification process with the caustic system was much faster and produced quite stable emulsions.
      With the surfactant-cosurfactant system, it has been observed (6) that the best oil displacement efficiency is achieved when the surfactant system spontaneously emulsifies with the oil, followed by rapid coalescence of the emulsified oil droplets (2).
      For the caustic systems used for recovering heavy oils it has been observed that the formation of stable emulsions by spontaneous emulsification is desirable (10, 11). The stable emulsions formed during caustic flooding tend to lower injected water mobility, viscous fingering, and water channeling while improving the sweep efficiency of the injected fluids. The produced fluids from this recovery technique are emulsions and must be demulsified.

Kinetics of Coalescence for Caustic Systems

      A very careful study of the kinetics of coalescence of an oil/water emulsion using photomicrographic droplet size analysis was done by Lawrence and Mills (17). They prepared their emulsions by homogenization. Their technique was recently modified by us to determine emulsion stability in petroleum sulfonate systems of interest in chemically enhanced oil recovery processes. These observations are given elsewhere (6, 7).
      The objective of the experiments presented here is to investigate effects of sodium hydroxide and sodium chloride on emulsion stability, and to establish the dynamics of spontaneous emulsification in a caustic system.
      Emulsions of crude oil in aqueous alkaline solutions were prepared by a hand homogenizer. Immediately after homogenization a representative sample of the emulsion was placed in a hemocytometer and photographed through the microscope. The hemocytometer had a specified grid volume of 10^{-4}mls. It was designed for a magnification of approximately 400 (i.e. 40X objective). The chamber of the hemocytometer did not apply a pressure on the emulsion and it prevented the emulsion from flowing in any direction.
      The Nikon LKe microscope with interference-phase-contrast was used to observe the emulsion. The Koehler illumination and the phase plates were aligned with a centering telescopic lens. The retardation of the light waves was adjusted with a 1/4 wavelength plate and the analyzer to develop the desired interference colors, thereby improving the contrast between the oil and aqueous phases (as seen through the microscope). A Bausch and Lomb stage micrometer was used to standardize the magnification of the microscope. The plain photomicrography was done with

Kodak photomicrography film No. 2483.

The coalescence of oil droplets in the emulsion was followed by photographing the emulsion at various intervals of time. The time interval between exposures was set according to the stability of the emulsion. For emulsions which took more than one hour to appreciably coalesce, samples of the emulsion were periodically placed between a glass slide and a cover slip and photographed.

Figure 3 shows a representative photomicrograph of the Huntington Beach crude oil-alkaline emulsion taken at 400 magnification in a continuous aqueous phase. The emulsion contained one part by volume of oil phase to twenty parts aqueous phase. This photograph is for a non-equilibrated system containing 0.06% NaOH and after the emulsion had been formed for fifty hours. The drop size distribution varies from about 3 to 20 microns.

The size distribution of the oil droplets in an emulsion was determined by counting and sizing the droplets from enlarged micrographs of the sample. The sizing was done from the micrographs taken at different time intervals after emulsion preparation. Figure 4 represents the cumulative percentage for different sizes on log-probability plot. The straight line behavior of the two samples indicates that the initial size distribution of the oil droplets can be represented by a log-normal distribution function. The vertical distance between the two straight lines is a representation of the faster coalescence of droplets in the sample which contained 0.06% NaOH. It should be noted that the log-probability plots of drop size distributions for the two samples exhibited similar behavior at higher times also.

Figure 5 shows the kinetics of coalescence for the caustic (0.05M NaOH, 1.0% NaCl), Thums Long Beach (heavy) crude oil system, with and without the co-surfactant n-hexanol (0.5%). This data shows that the mean droplet volume (which is proportional to 1/number of droplets) increases with time. The addition of hexanol alters the kinetics of interdroplet coalescence to a level that the emulsion almost totally coalesces within ten days.

Reisburg and Doscher (18) reported that the suppression of a semi-solid film formation at the oil-water interface played an important role in improving oil recovery by caustic flooding. Our recent experiments with the co-surfactant n-hexanol have shown that it reduces the thickness of the surfactant film and enhances the rates of inter-droplet coalescence in surfactant systems (6). Figure 5 shows that our findings with the petroleum sulfonate systems could be extended to alkaline systems.

A series of experiments were conducted to determine the emulsion stability in caustic systems. Figures 6 and 7 show data for the kinetics of coalescence and hence emulsion stability for the crude oil from Huntington Beach (Lower Main Zone), California (oil gravity of 23°API and oil acid number of 0.65). Figure 6 shows data for a nonequilibrated system and for a very low concentration of NaOH (0.003%) and 1% NaCl. This emulsion is unstable. Figure 7 shows data for two different concentrations of

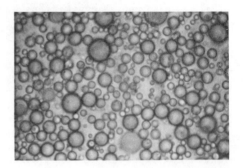

Figure 3. Microphotograph of Huntington Beach crude oil–caustic emulsion

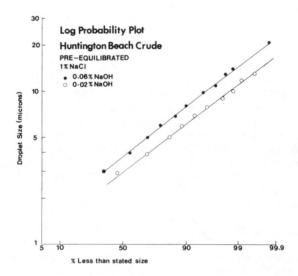

Figure 4. Drop-size distributions for Huntington Beach crude oil–caustic emulsions

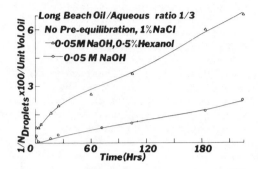

Figure 5. *Kinetics of coalescence for nonequilibrated sample of Long Beach crude oil–caustic system*

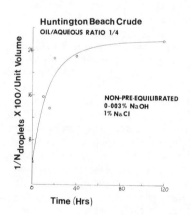

Figure 6. *Kinetics of interdroplet coalescence for caustic system*

NaOH(0.02% and 0.06%) and 1% NaCl. These data are for equilibrated samples which have been pre-equilibrated for twelve hours.
There is a pronounced effect of caustic concentration on emulsion
characteristics as seen from this figure.

Effect of Interfacial Properties on Emulsion Stability

One of the main objectives of this study has been to determine the effect of interfacial properties on coalescence,
emulsion stability and oil recovery efficiency for various surfactant and caustic systems. We have recently reported (6, 19)
that for a petroleum sulfonate system there is no direct correlation between rates of coalescence and interfacial tension or
interfacial charge. However, a qualitative correlation has been
found between coalescence rates and interfacial viscosities.

The interfacial tension for the crude oil-caustic system was
measured by the spinning drop technique. The instrument used is
similar in design to the one reported by Schechter and Wade (2).

We measured the electrophoretic mobilities of crude oil
droplets in alkaline solution using a Zeta Meter (20). Since the
droplet sizes were larger than one micron, the zeta potentials
were calculated from electrophoretic mobilities using
Smoluchowski's formula.

Figure 8 shows the values of electrophoretic mobility and
interfacial tension as a function of the NaOH concentration for
the Long Beach crude oil which has been equilibrated with the
alkaline solution. This figure shows that the electrophoretic
mobility increases and then decreases with increasing caustic
concentration. It should be noted that the maximum in electrophoretic mobility appears to correspond to a minimum in interfacial tension. This finding is consistent with our recent
results for surfactant systems (6) and those of Shah and Walker
(21).

The interfacial tension values reported for the caustic system in Figure 8 are comparable to the values reported recently
in reference (22). Our experiments which have been conducted
at a room temperature of about 25°C show that 0.1 to 0.4 weight
percent concentrations of NaOH and 1.00 weight percent NaCl can
lower the interfacial tension between the aqueous solution and
the crude oil substantially below a value of 0.01 dynes/cm or
that required for emulsification. We have previously discussed
the stability of these emulsions (Fig. 5). In the experiments
run on fired Berea cores, it was reported that a concentration
of 0.1% NaOH and 1% NaCl in the caustic crude oil system resulted in a drastic reduction in residual oil saturation. The
details of these tests are given in reference (22).

Figure 9 shows the interfacial tension between the
Huntington Beach crude oil and sodium hydroxide as a function of
the age of the interface. The interfacial tension is found to
increase with the age of the interface. This behavior is similar

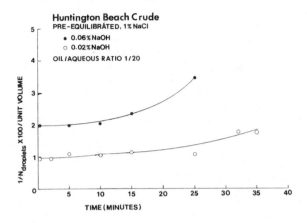

Figure 7. *Kinetics of coalescence for equilibrated samples of Huntington Beach crude oil–caustic systems*

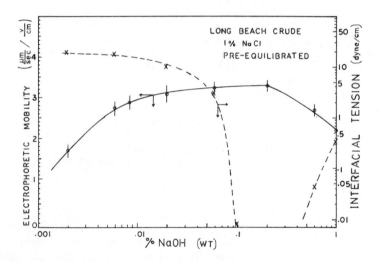

Figure 8. *Effect of caustic concentration on electrophoretic mobility and inter-facial tension of an equilibrated sample of Long Beach crude*

to the one reported by McCaffery (23) for the heavy crude oils
contacted with caustic solutions. It has been suggested that
this behavior is due to the migration away from the oil/water
interface of the sodium soap products, which are formed when
acidic compounds in the crude oil react at the interface with
sodium ions in the aqueous phase.

Figures 10 and 11 display both the interfacial tension and
electrophoretic mobility data for Huntington Beach crude as a
function of NaOH concentration in aqueous solutions containing
1% NaCl. It is to be noted that as in the case of the Long Beach
crude, the minimum in interfacial tension corresponds to the
maximum in electrophoretic mobility and hence interfacial charge.
It should be further stated that this behavior is the same for
both the equilibrated system (Figure 10) and for the non-
equilibrated system (Figure 11). In the latter case, the interfacial tension reported here is the initial value at a given
NaOH concentration. It is observed that the minimum interfacial
tension for the pre-equilibrated system lies at about 0.06% NaOH
whereas the minimum for the non-equilibrated sample is between
0.003% and 0.005% NaOH concentration. Furthermore, the minimum
interfacial tension in this case is below 0.001 dynes/cm.

Figure 12 shows plots of interfacial tension and electrophoretic mobility for samples containing no sodium chloride for
both equilibrated and non-equilibrated systems. It is interesting to note that the electrophoretic mobility values are about
the same for both these systems over a wide range of concentrations of NaOH. However, the interfacial tension behavior is
quite different from that expected. It should be noted that the
maximum in electrophoretic mobility does not correspond to the
minimum in interfacial tension. In addition, the electrophoretic
mobility values are much higher for those samples containing no
sodium chloride compared to those which contained NaCl (Figures
10 and 11). The minimum interfacial tension occurs at much
lower NaOH concentration for samples containing NaCl.

Figure 13 exhibits both interfacial tension and electrophoretic mobility for the Huntington Beach Field crude oil
against sodium orthosilicate containing no sodium chloride. The
interfacial tension values are observed to be higher for the
non-equilibrated sample in this case than for the caustic system
reported in Figure 12. The minimum interfacial tension of 0.01
dynes/cm occurs at about 0.2% sodium silicate as opposed to a
value of less than 0.002 dyne/cm at about 0.06% NaOH. It is
interesting to note, however, that the maximum electrophoretic
mobility is the same for the two systems. Once again, it should
be noted that a maximum in electrophoretic mobility does not
correspond to a minimum in interfacial tension for those samples
which contained no sodium chloride.

In an attempt to evaluate the effect of electrostatic interaction on the stability of crude oil-caustic or orthosilicate
emulsions, the total interaction energy (V_t) between two oil

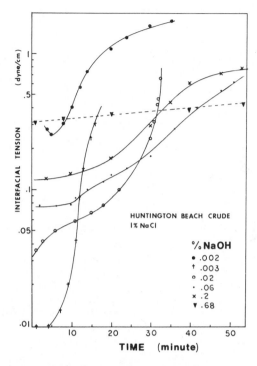

Figure 9.   Effect of the interface age on interfacial tension

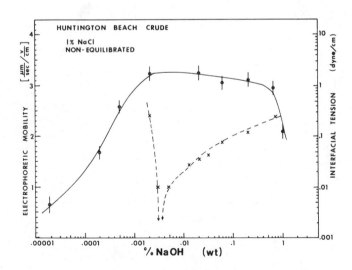

Figure 10.   Effect of caustic concentration on electrophoretic mobility and inter-facial tension of nonequilibrated sample of Huntington Beach crude

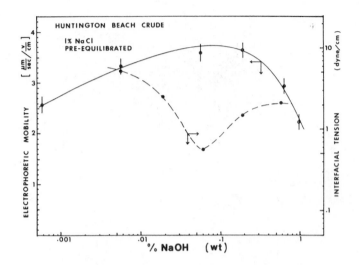

Figure 11. *Effect of caustic concentration on electrophoretic mobility and inter-facial tension of equilibrated sample of Huntington Beach crude*

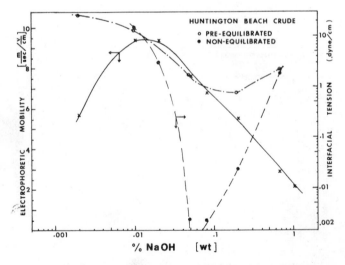

Figure 12. *Effect of caustic concentration on electrophoretic mobility and inter-facial tension with no NaCl*

droplets has been calculated using Verway-Overbeek theory (24).
In such a calculation, the Hamaker constant is assumed to be
10^{-13} erg; oil droplet size is taken as 1 micron and zeta potential is -50mv for 0.06% NaOH with 1% NaCl and -96.5 mV for 0.06%
NaOH with no sodium chloride. Figure 14 shows plots of V_t vs. H,
the closest distance between the surfaces of the two droplets.
At 1% NaCl, there is a monotonic increase in attractive potential
energy, therefore, oil drop flocculation is irreversible and oil
droplets can be brought together at contact distances. For the
system with no sodium chloride, there is a repulsive energy
barrier between oil droplets at interparticle distances below
600A°, therefore, oil-drop flocculation is reversible, and hence
conditions are extremely unfavorable for coalescence. Thus the
emulsion is extremely stable.

We have attempted to relate the emulsion stability test
results as reported for caustic systems in Figure 7 with interfacial properties to gain a better understanding of the importance
of these properties in alkaline water flooding. Table 1 summarizes the values of interfacial tension, electrophoretic
mobility and zeta potential, and interfacial shear viscosity for
equilibrated samples of Huntington Beach crude oil against 0.02%
and 0.06% NaOH containing 1% NaCl. The interfacial viscosities
for crude oil-caustic solutions were measured using the deep-
channel viscous traction interfacial viscometer according to the
new method outlined by Wasan et al. (25, 26). This table clearly
shows that an emulsion of oil in 1.0% NaCl and 0.02% NaOH with
an interfacial tension of 2.6 dyne/cm and zeta potential of -45mv
produced more stable emulsions than in 1.0% NaCl and 0.06% NaOH
with an interfacial tension of 0.6 dyne/cm and zeta potential of
about -50mv. It is most interesting to note that the sample with
0.02% NaOH has an interfacial viscosity value of 8.04×10^{-2} sp.
as compared to 5.83×10^{-2} sp. for the 0.06% NaOH sample; hence,
the higher the interfacial viscosity, the greater the emulsion
stability. This finding for the caustic systems is similar to
our recent results for emulsions of crude oil in petroleum sulfonate solutions as mentioned earlier in this paper. We made no
attempt to correlate the emulsion stability results for non-
equilibrated caustic samples with interfacial viscosities as
these have yet to be measured.

Micellar Aggregate Distribution

Micelle size and structure are stabilized by surfactant
interactions and bonding. Therefore additives that destabilize
micellar structure also disrupt the interactions and bonding of
the surfactants adsorbed at the oil-aqueous interface. The disruption of the surfactant interactions and bonding at the interface leads to a weakening of the interfacial film and thereby
promotes coalescence. The micellar aggregate size distributions
for surfactant systems under consideration for chemical flooding,

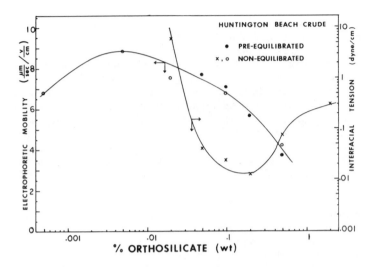

Figure 13.   *Effect of orthosilicate on electrophoretic mobility and interfacial tension with no NaCl*

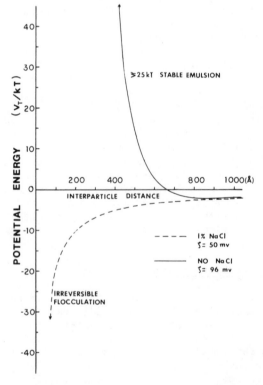

Figure 14.   *Energy of interaction vs. separation distance between two droplets*

Table I:
Interfacial Properties for the Crude Oil-Caustic System

Oil	Aqueous Phase	Initial Inter-facial Tension dynes/cm	Electrophoretic Mobility $\frac{\mu m}{sec}$ /v/cm	Zeta Potential mv	Interfacial Viscosity sp.
Huntington Beach Crude Lower Main Zone S-47	0.02% NaOH 1% NaCl	2.6	3.5	-45	8.04×10^{-2} sp.
	0.06% NaOH 1% NaCl	0.6	3.8	-50	5.83×10^{-2} sp.

to our knowledge have not been reported in literature (6, 7, 27).
The determination of micellar aggregate size distributions for
three phase systems could be used to understand the interactions
which make these systems desirable for enhanced oil recovery.
The micellar aggregate size distribution is also important because large micellar aggregates plug reservoir pores with small
throat diameters (28). This plugging may cause poor oil recovery.

The objectives of this study was to determine the changes in
micellar aggregate size distributions caused by oil/water ratio,
co-surfactant and three phase development in petroleum sulfonate
systems and by a co-surfactant in a caustic system.

In the study of micellar aggregate size and structure attention was focused on the large micelles of various surfactant
systems. Mukerjee (29) calls large micelles "association
colloids" because they result from the association of small
molecules.

A Model TA-II Coulter Counter with the population accessory
was used to measure size distributions of micellar aggregates.
The aperture tube with the 30 micron orifice was attached to the
counter. A 1.5% NaCl solution was filtered with a 0.22 micron
Millipore filter. The Coulter Counter was flushed with this
solution. A sampling time of five seconds (equivalent to 0.05
mls) was selected. The channels of the counter were standardized
with 2.02 micron polymer beads. The orifice of the aperture tube
was examined by the microscope attached to the counter and by
the oscilloscope. The micellar aggregate size distributions for
the aqueous samples were determined in channels 2 through 15 by
differential counting in each channel. The micellar aggregate
size distributions determined by the Coulter Counter were
verified by the Nikon LKe transmitted light microscope. The
magnification of 1,000 (with oil immersion) was used along with
a green light filter to improve resolution. To prove that the
"particles" being sized by the Coulter Counter were surfactant
micellar aggregates and not oil droplets the interference-phase
contrast technique and ultra-sonication were applied.

The phase difference between he "particles" (which were
detected by the Coulter Counter) and oil droplets of the same
size was determined. This phase difference was large. This
meant that the refractive index of the "particles" was different than the refractive index of the crude. The "particles"
therefore were not oil droplets. This only leaves the possibility that they were micellar aggregates of surfactant. However,
it is possible that there was some crude oil dissolved in the
hydrophobic cores of the micelles that constitute these aggregates.

The presence of large micellar aggregates was also verified
as follows. The particles were disrupted by sonic energy using
an ultra-sonicator (Labline, Inc.). The decrease in the micellar
aggregate sizes due to sonication was monitored by the Coulter
Counter. However, within a period of two days the "particles"

had grown back to their original size distribution. The refor-
mation of these "particles" can not be explained if they were
considered to be oil droplets.

Szabo (28) has reported that the plugging of pores by surfactants could be partially alleviated by the addition of an
alcohol (i.e. co-surfactant). This would indicate that the co-
surfactant disrupts the structure of the surfactant micelles.

It has been shown in our recent publications (6, 7, 27)
that the addition of a co-surfactant greatly enhances the coalescence rates of oil droplets. The co-surfactant must have disrupted the surfactant interactions at the oil-aqueous interface.
We have also previously reported the significant changes in
micellar aggregate size distribution caused by the equilibration
of the aqueous surfactant solution against the crude oil.

The equilibrium phase behavior of systems of oil and water
containing appreciable amounts of surfactant is characterized by
the presence of microemulsions. The microemulsions are stable
oil-water dispersions caused by the incorporation of amphipathic
molecules (surfactants and co-surfactants) in the oil and water
phases.

When an aqueous system containing a surfactant, co-
surfactant and of intermediate salinities is allowed to equili-
brate with crude oil, the mixture sometimes separates into three
phases. One of these phases is the aqueous phase which contains
very little surfactant. This is called the lower phase. The
second phase is called the middle phase. This phase is a microemulsion which contains large amounts of both oil and water and
nearly all the surfactant. The third (upper) phase contains the
oil. Systems of oil and aqueous phases which show this phase
behavior are said to exist in the "beta" region of the phase
diagram. The "beta" type systems have been shown to form the
least stable emulsions and thereby result in enhanced oil recovery (30).

The "beta" system developed in the 3% Petrostep 420 + 1.5%
NaCl + 0.58% hexanol system upon equilibration of the aqueous
phase with Salem crude oil at an oil/aqueous ratio of 1/4. The
Coulter Counter and microscopy techniques mentioned earlier were
employed to characterize this three-phase formation (i.e. a
shift to the "beta" region).

Figure 15 compares the micellar aggregate size distribution
for the aqueous phase containing 3% Petrostep 420, 1.5% NaCl and
0.58% hexanol with the size distribution of micellar aggregates
in the middle and lower phases which developed when the aqueous
phase was equilibrated with crude oil. This figure shows that
most of the surfactant is contained in the midele phase after
equilibration. The middle phase was examined by interference
phase contrast microscopy. It was observed that the middle
phase consisted of an oil-in-water emulsion. The size of the
oil droplets in the emulsion ranged from 2 to 3 microns in diameter. The oil droplets were encapsulated in a thick film of

surfactant.

The oil/aqueous ratio was found to be important for three phase development and disruption. Three phases were formed for the Petrostep system at an oil/aqueous ratio of 1/4, but the system had only two phases when the ratio was increased to 1/2 (27). This behavior indicates the very complex nature of the oil/aqueous interactions which also must be understood in order to analyze the processes important for enhanced oil recovery by surfactant systems.

The very low interfacial tensions reported for many crude oil-caustic systems should permit substantial reduction of residual oil saturation by the mobilization of trapped oil. We have discussed earlier that crude oil-caustic tension is low initially where reactants meet at a fresh interface but the interfacial tension increases as reaction products diffuse into the bulk phases. The technique of determining micellar aggregate size distributions could be used to study the diffusion of the reaction products into the aqueous phase.

The reason for understanding the effect of n-hexanol on the micellar aggregate size distribution in the caustic systems are similar to those for the sulfonate systems presented earlier.

Figure 16 shows that the presence of the co-surfactant hexanol resulted in the formation of much smaller micellar aggregates of the natural surfactant than those formed in the system which did not contain hexanol. The caustic systems contained 0.05 M NaOH (0.2% by weight) and 1% NaCl. These observations are similar to those made for the sulfonate system (19, 27).

Surfactant Film Characterization

Many crude oils contain surface active compounds that will collect at an oil-water interface to form a rigid film. Since interfacial films are important in the displacement of oil by water in reservoir rock, the presence of rigid films in the reservoir could alter the recoverable oil by water invasion (31).

When the aqueous phase contains a surface active compound, the compound collects at an oil-water interface to form a surfactant film. This film may or may not be rigid. The physical properties of this film affect emulsion stability and thereby influence the yield of crude oil by surfactant floods.

The determination of the molecular orientation and the thickness of this surfactant film is very important in understanding the basic factors which affect emulsion stability (32).

There are various types of molecular orientations possible in the surfactant film. The surfactant molecules may be in the form of crystals, isotropic gels or one or more of several different types of mesophases (33). The techniques which may be used to study the molecular orientations inside these surfactant films have been reviewed by Singer (34).

Phase contrast microscopy at a magnification of 1000 was

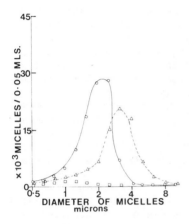

Figure 15. Micellar aggregate distributions for Petrostep 420 with n-hexanol, three-phase system. For the aqueous phase containing 3.0% P. 420 + 1.5% NaCl + 0.58% hexanol, nonequilibrated (○), and equilibrated for a Salem crude oil/aqueous ratio of 1/4, Middle Phase (△) and Lower Phase (□).

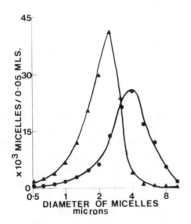

Figure 16. Natural crude oil surfactant micellar aggregate size distributions for Long Beach crude/caustic system. The aqueous phase containing 0.05M NaOH without hexanol (●) and with 0.50% hexanol (▲).

used to qualitatively examine the rigidity of the films for the
equilibrated 3% Petrostep 420+n-hexanol system. This system
formed very flexible films which were observed to rupture readily
upon contact of two oil droplets (6, 7).

The formation of surfactant crystals (i.e. liquid crystals)
at the oil-aqueous interface can be easily determined by the
use of polarized light microscopy (35).

Liquid crystals have been observed in the surfactant (20,
27) films surrounding oil droplets for the TRS 10-80 system
employed by Strange and Talash (3) and Whiteley, et al. (4, 5)
for the Salem low tension water flood field tests. The crystals
were visible through crossed polarizing plates at a magnification
of 400. The presence of crystals in these films indicates that
the surfactant molecules are highly organized for this system.
This would tend to make the interface rigid and thereby lower
coalescence rates. The TRS 10-80 system had very poor coalescence as reported previously by us (6).

Liquid crystals were also observed in the interfacial films
of water-in-oil emulsions present in Thums Long Beach crude oil
produced by secondary oil recovery.

Fig. 17 shows a photograph of a crude oil droplet dispersed
in non-equilibrated aqueous phase consisting of .02% NaOH and 1%
NaCl. The droplet is approximately 16 microns in diameter.
The film around the droplet appears to be thicker than those
observed in the case of petroleum sulfonate systems (6). The
emulsion was found to be more stable.

Conclusions

(1) The mechanisms of a spontaneous emulsification in
petroleum sulfonate and caustic systems have been described.
The differences in behavior in surfactant and alkaline solutions
have been characterized by high speed cinephotomicrography.

(2) Preliminary results on the kinetics of coalescence of
both the Long Beach and the Huntington Beach crude oil droplets
in caustic systems have been presented.

(3) The effects of interfacial tension, interfacial charge
and interfacial viscosity on coalescence, and emulsion stability
for crude oil emulsions in alkaline solutions have been assessed.
It was observed that the NaOH concentration which yields higher
interfacial shear viscosity also results in higher emulsion
stability.

(4) Further data on the effect of interfacial viscosity on
emulsion stability and its subsequent effect on oil recovery
efficiency by alkaline water flooding are needed.

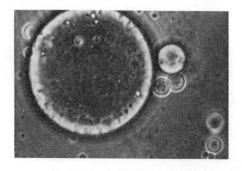

Figure 17. Interference phase contrast micrograph of crude oil in caustic solution

(5) The maximum electrophoretic mobility or zeta potential corresponds to the minimum interfacial tension for the caustic systems containing 1% NaCl but this correlation is not valid for systems which do not contain NaCl.

(6) The interfacial tension between the crude oil and the alkaline solution is a strong function of the age of the interface. The interfacial tension increases with the age of the interface.

(7) The interference phase contrast technique combined with high resolution optical sectioning needs to be developed further to measure the thickness of the film surrounding oil droplets in caustic systems. The film thickness and the molecular packing in the film need to be correlated with the stability of an emulsion system. Preliminary results obtained to date are quite encouraging.

(8) Micellar aggregates with relatively large diameters have been found for both the petroleum sulfonate and the caustic flooding systems. The role of these micellar aggregates in enhanced oil recovery by chemical flooding needs to be established.

Abstract

In this paper we present microscopic observations and high speed cinephotomicrographic examinations of spontaneous emulsification phenomena in both surfactant and caustic systems. We also report results of laboratory experiments involving acidic crude oils from both Thums Long Beach (Wilmington field) and Huntington Beach containing natural surface-active agents which greatly influence interfacial rheological properties and emulsion stability. The crude oils are contacted with caustic solutions containing electrolytes and alcohols. Data on interfacial viscosities, interfacial tension, interfacial charge, micellar aggregate distributions, and the microscopic observations of inter-droplet coalescence of oil-in-water emulsions are reported. Photomicrographic observations of the rate of coalescence-rupture processes with droplet size determinations provide an insight into the stability of emulsion systems.

Acknowledgment

The financial support provided by the National Science Foundation and Department of Energy is acknowledged.

Literature Cited

1. Childress, G. S., "A Microvisual Study of the Initial Displacement of Residual Hydrocarbons by Aqueous Surfactant Solutions," M. S. Thesis, University of Texas at Austin (December, 1975).

2. Schechter, R. S. and Wade, W. H., "Spontaneous Emulsification and Oil Recovery," Report Submitted to ERDA under Grant No. EY-76-S-0031, November, 1977.
3. Strange, L. K., and Talash, A. W., SPE Paper No. 5885 presented at the SPE-AIME Fourth Symposium on Improved Oil Recovery, Tulsa, March 22-24, 1976.
4. Whiteley, R. C. and J. W. Ware, SPE Paper No. 5832 presented at the SPE-AIME Fourth Symposium on Improved Oil Recovery, Tulsa, March 22-24, 1976;-also in J. Pet. Tech., (1977), 925.
5. Widmyer, R. J., Salter, A., Frazier, G. D., and Graves, R. H., J. Pet. Tech., (1977), 933.
6. Wasan, D. T., Shah, S. M., Aderangi, N., Chan, M. and McNamara, J. J., "Observations on the Coalescence Behavior of Oil Droplets and Emulsion Stability in Enhanced Oil Recovery," Paper SPE 6846, presented at the SPE-AIME 52nd Annual Fall Technical Conference, Denver, Colorado, October 9-12, 1977, paper accepted for publication in Soc. Pet. Eng. J.
7. Wasan, D. T., Shah, S. M., Aderangi, N., Chan, M., McNamara, J. J., Paolicchi, R., Patel, P., and Mohan, V., Paper No. B-4 presented at the Third Annual Symposium on Enhanced Oil, Gas Recovery and Improved Drilling Methods, Tulsa, August 30 - September 1, 1977.
8. Cooke, C. E., William, R. E. and Kolodzie, P. A., J. Pet. Tech. 26, 1365 (1974).
9. Ehrlich, R., Hasiba, H. H. and Raimondi, P., J. Pet. Tech., (1974), 1335.
10. Jennings, H. Y., Johnson, C. E., and McAuliffe, C. D., J. Pet. Tech., (1974), 1344.
11. Jennings, H. Y., Soc. Pet. Tech. J., (1975), 197.
12. Johnson, C. E., J. Pet. Tech., (1976), 85.
13. Radke, C. J. and Somerton, W. H., Paper No. B-5 presented at the Third Annual ERDA Conference on Enhanced Oil, Gas Recovery and Improved Drilling Methods, Tulsa, August 30 - September 1, 1977.
14. Sarem, A. M. SPE Paper No. 4901 presented at the California Regional Meeting, San Francisco, April 4-5, 1974.
15. Wagner, D. R. and Leach, R. O., Trans. AIME, (1959) 216, 65.
16. Gopal, E. S. R. "Principles of Emulsion Formation," in P. A. Sherman (ed.) "Emulsion Science" p. 60 Academic Press, New York, 1968.
17. Lawrence, A. S. V. and Mills, O. S., Discussion, Faraday Society (1954) 18, 98.
18. Reisburg, J., and Doscher, T. M., Prod. Monthly, (1956) 20, 43.
19. Wasan, D. T., Chan, M., Shah, S. M., Sampath, K., and Shah, R. "Interrelationship of Emulsion Stability and Interfacial Viscosity in Improved Oil Recovery," paper presented at the Engineering Foundation Conference on Theory, Practice

and Process Principles for Physical Separation, Asilomar,
California, October 30 - November 4, 1977.

20. Chan, M., M.S. Thesis in preparation, Illinois Institute of
    Technology, Chicago, IL (1978).

21. Shah, D. O., and Walker, R. D., Second Semiannual Report,
    University of Florida, Project C33 and Project C27, June,
    1976.

22. City of Long Beach and Thums Long Beach Co., "Caustic Water-
    flooding Demonstration Project, Ranger Zone, Long Beach
    Unit, Wilmington Field, California," Annual Report sub-
    mitted to U.S. ERDA, Contract No. EF 77-C-03-1396, June,
    1977.

23. McCaffery, F. G., J. Can. Pet. Tech., (1976), 71.

24. Kitchner, J. A., and Mussellwhite, P. R., in "Emulsion
    Science" (ed. P. Sherman), Chapter 2, Academic Press,
    London and New York, 1968.

25. Wasan, D. T., Gupta, L., and Vora, M. K. AIChE J. (1971) 17.

26. Mohan, V. and Wasan, D. T., "A New Method for the Measurement
    of Shear Viscoelasticity at Liquid-Liquid Interfaces Con-
    taining Surfactants and Macromolecules," Colloid and
    Interface Sci., Vol. IV, 439-446, Academic Press, New York,
    1976 . Editor M. Kerker.

27. Wasan, D. T., "An Overview of Interfacial Phenomena in En-
    hanced Oil Recovery System," paper presented at the Fall
    Meeting of the Society of Rheology, October 24-27 1977,
    paper accepted for publication in J. Rheology,(1978).

28. Szabo, M. T. 1977, "Micellar Shear Degradation, Formation
    Plugging and Inaccessible Pore Volume," SPE paper No.
    6772 presented at the 52nd Annual Technical Conference of
    the SPE-AIME. Denver, Co., October, 1976.

29. Mukerjee, P., J. Pharm. Sci., (1974) 63, 972.

30. Vinatieri, J. E., SPE Paper No. 6675, New Paper submitted to
    J. Pet. Tech.,(1977).

31. Bourgoyne, A. T., Ph.D. Thesis, University of Texas, Austin,
    Texas (1970).

32. Krog, N., "Structures of Emulsifier-Water Mesophases Related
    to Emulsion Stability," Felte Scifen Austrichm (1975), 77,
    (7), 267. (English translation).

33. Wennerstrom, II, Persson, N. and Lindman, B., "Deuteron NMR
    Studies on Soap-Water Mesophases," in K. L. Mittal (ed.),
    "Colloidal Dispersions and Micellar Behavior," ACS Sym-
    posium Series 9, ACS, Washington, D. C., 1975.

34. Singer, S. J. "The Molecular Organization of Biological Mem-
    branes," In R. I. Rothfield (ed.), Structure and Function
    of Biological Membranes., Academic Press, New York, 1971.

35. Hartshorne, "The Microscopy of Liquid Crystals," McCrone
    Publications, London, England, 1973.

RECEIVED November 6, 1978.

Comparison of Solution Properties of Mobility Control Polymers

ESREF UNSAL, JOHN LARRY DUDA, and ELMER KLAUS

Department of Chemical Engineering, The Pennsylvania State University, University Park, PA 16802

Showing the technical feasibility of polymer flooding or testing a polymer for mobility control in an oil reservoir requires five steps. The first step is to determine the solution properties of this polymer. This investigation which can be carried on in a laboratory is the most inexpensive step and it can also produce the most fundamental data. The second step in the process of evaluating a polymer is to determine the injectability of a polymer solution. Tight reservoirs require the solutions to be injectable without leading to plugging. Formations with permeabilities in excess of 200 or 300 md may not require this step. The final three steps all involve actual porous media tests and they tend to be more tedious and more expensive, nevertheless essential: preliminary tests with inexpensive porous media that simulate the actual reservoir rock (filter papers, metal porous media, parallel capillaries, etc.), tests with cores in the laboratory, and field tests. The more understanding that is obtained in the first three steps, the less of the latter two steps that have to be carried out. It is desirable to develop methods to correlate the information obtained in the first three laboratory tests with the results of the latter tests in order to minimize the number of expensive experiments required for the design of a flood.

This paper will address itself exclusively to the first of the above mentioned steps, namely, the solution properties of mobility control polymers. The polymers investigated cover quite a spectrum: Kelzan®- XC (xanthan gum), Pusher®-700, Colloid®XH$_0$ and XH3, and Natrosol® 250 HHR. The brief descriptions of these polymers are given below:

a. Pusher®: This is a partially hydrolyzed polyacrylamide which has been the most widely employed polymer in enhanced oil recovery applications. It has some undesired characteristics such as vulnerability to mechanical degradation and sensitivity to salt.

b. Xanthan gum: This is a biopolymer which also has been

0-8412-0477-2/79/47-091-141$07.50/0

utilized widely even though biodegradation and injectivity to
tight oil sands have been sources of concern. Xanthan gum is the
exocellular biopolysaccharide produced in a fermentation process
by the microorganism Xanthomonas campestris (1). Three different
monosaccharides have been found in xanthan gum: mannose, glucose,
and glucuronic acid (2). The main chain of xanthan gum has B-D-
glucose units linked through the 1- and 4- positions (1).

c. Colloid®: This polymer is a polysaccharide produced by
conventional chemical treatments as opposed to fermentation. It
is a combination of sodium cellulose sulfate esters in which the
substituents are uniformly distributed over the macromolecule (3).
Many properties of this polymer are similar to the properties of
other polysaccharides.

d. Natrosol®: This is a hydroxy ethyl cellulose derived
from cellulose with a nonionic character (4). It will be shown
that this polymer provides some advantages over the three polymers
mentioned above.

The literature on mobility control polymers includes many
over-simplifications concerning the behavior of these polymer
solutions such as: polysaccharides are insensitive to salt, and
polyacrylamides easily degrade in a shear field. This study will
show that the differences between solutions are a matter of degree and the basic behaviors and the mechanisms are common to all
mobility control polymers.

The block of data concerned with this area generally does not
consider the developments concerning the solution behavior of non-
mobility control polymers which has been reported in the polymer
literature. Implicitly, one is led to believe that the mobility
control polymer solutions are unique and have little in common
with more conventional polymer solutions. This study will show
that there is a direct qualitative correlation between the behavior of mobility control polymer solutions and other solutions
of macromolecules. In particular, it is demonstrated that the
viscous properties are directly related to the hydrodynamic size
of the polymer chain; and the influence of system characteristics
such as salt concentration, shear rate, etc., can be correlated
with the effective size of the polymer molecule in solution. Consequently, this study suggests that more emphasis should be placed
on the measurement of the molecular size of mobility control polymers in solution if a fundamental understanding of these solutions
is to be developed.

Molecular Size in Solution

The dominate factor which controls the solution properties of
mobility control polymers is the configuration (and hence size)
that the molecule assumes in a given environment. Although the
viscosity of a polymer solution is related to the hydrodynamic
size of the polymer molecules, it is difficult to determine the
unique relationship between viscosity and size. However, exten-

sive studies indicate that the solution property referred to as
intrinsic viscosity is related to size in a simple universal manner (5). This discussion will focus on how to determine intrinsic viscosities and molecular sizes for the polymers being investigated.

The intrinsic viscosity [ ] indicates how the viscosity
changes with polymer concentration at infinite dilution:

$$[\eta] = \lim_{C_p \to 0} \left| \left[\frac{\eta}{\eta_s} - 1 \right] / C_p \right| \tag{1}$$

C_p = polymer concentration, g/l

η = viscosity, cp

η_s = viscosity of solvent, cp

$[\eta]$ = intrinsic viscosity, 1/g

The intrinsic viscosities are obtained by making viscosity measurements at different polymer concentrations and by plotting the
above expression against the concentration. The limit that this
quantity assumes as the infinite dilution is approached is the
value desired. Polymer solutions exhibit a Newtonian behavior at
low shear rates, changing to a non-Newtonian flow at higher shear
rates (6). The intrinsic viscosities should be determined at the
low shear rate range. The most commonly used equation that relates the intrinsic viscosity and the molecular weight of a macromolecule is the Mark-Houwink equation:

$$[\eta] = K_I M^{\alpha} \tag{2}$$

The constants K_I and α can be either measured by an overlap of intrinsic viscosity data and another technique of measuring molecular weights or can be estimated from the considerations of molecular configurations.

The molecular weights and the sizes of the polymers investigated in a 200 ppm NaCl solution at 25°C are given in Table 1.
The values for xanthan gum were also reported in an earlier work
(7). The molecular sizes were obtained by using the Flory relation (8). There are alternate discussions as to what the configuration and the size of xanthan gum molecules in solution are.
Whitcomb, Ek, and Macosko have presented an interpretation of the
intrinsic viscosity data assuming a cylindrical rod conformation
(9). The K_I and α values for Pusher®are given by Lynch and Mac-
Williams (10). It should be noted that a range of K_I and α values
for polyacrylamides can be found in the literature (11).

The intrinsic viscosity of Colloid® is close to the value
for xanthan gum. The K_I and α values are not available, hence
the molecular weight and size are not calculated. These numbers
are expected to be similar to that of xanthan gum. The actual

molecular weight of Colloid Ⓡ samples range from 1 to 2 million
based on the conversion of the reaction and the degree of substi-
tution (3). The K_I and α values for Natrosol   are obtained from
the polymer literature (12).

Table 1

Molecular Weights and Sizes of Polymers

Polymer	[η] (1/g)	K_I	α	Molec. Size (Å)	Molec. Wt. (x 10^{-6})
PusherⓇ -700	9.9	3.73x10^{-4}	0.66	2950	5.0
KelzanⓇ -XC	8.7	1.18x10^{-5}	0.85	3290	8.0
ColloidⓇ-XH$_0$	8.2	----	----	----	(1-2)
NatrosolⓇ-250 HHR	1.8	0.95x10^{-5}	0.87	1020	1.2

The intrinsic viscosity (hence the molecular size) of the
mobility control polymers is a strong function of the environment
or the electrolyte concentration of the solution and the temperature. The size is also affected by mechanical degradation. As
shown in Table 2, increasing the salt concentration leads to a
smaller size in polyacrylamides. This effect is more drastic at
low salt concentrations.

Table 2

Effect of NaCl on Molecular Size of PusherⓇ

NaCl Concentration (ppm)	Intrinsic Viscosity (1/g)	Size (Å)
200	9.90	2950
1000	7.18	2650
10000	6.42	2550

Preparation and Evaluation of Solutions

Mixing. Dissolving a polymer is a slow process which occurs
in two stages. First, a gel is produced when the solvent molecules start to solvate and associate with the polymer chains (5).
Then this gel disintegrates into a molecular dispersed solution.
Different polymers exhibit quite different characteristics during
this process, and this process is influenced by variables such as
shear rate, temperature, and salt concentration.
     Severe shear conditions are required to disperse the gels
formed by xanthan gum (7). Figures 1 and 2 show a comparison of

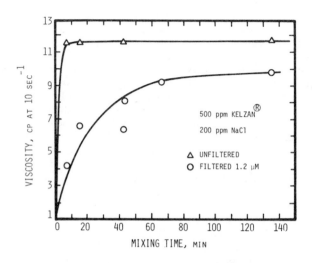

Figure 1. *Mixing in impeller system* (7)

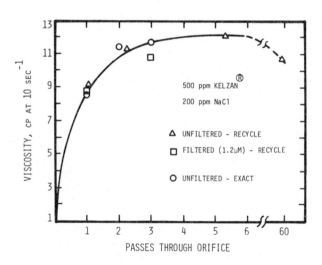

Figure 2. *Mixing in orifice blender* (7)

the dispersion of xanthan gum molecules in orifice and impeller
systems. These two devices represent two mixing modes. Stirring
with propeller is a mild means of dispersion without significant
mechanical degradation. Orifice blending represents severe mixing with different extents of downstream cavitation depending on
the size of the orifice. Solution viscosities are chosen as the
criterion for evaluation of mixing. Since all the polymer buffer
solutions exhibit non-Newtonian viscosity properties, the comparisons have to be made at a fixed shear rate. A shear rate of 10
sec^{-1} was chosen for comparison since this corresponds approximately to a typical chemical flood of 0.3 ft/day in a tight reservoir. In propeller blending, viscosity increases rapidly to a
high value as can be seen on Figure 1. Examining the quality of
dispersion by measuring the viscosities of unfiltered solutions
can be misleading. Most of the viscosity is lost after a 1.2 µM
filtration. This shows that the gels that form contribute to the
viscosity but they are filtered out as soon as the solution is
passed through a hole of size 1 or 2 µM. Severe shearing is required to dissolve these gels. The viscosities of filtered propeller solutions go up as mixing time is increased and they finally level off at long mixing times.

In orifice blending, the viscosity increases as the solution
passes through the orifice as shown on Figure 2. The viscosity
loss upon filtration is small in this case, indicating that this
high shear field device is required for the dispersion of xanthan
gum.

The number of passes required through an orifice to form a
well-dispersed solution depends on the polymer concentration. The
reservoir characteristics are needed for the final design of the
orifice mixing system. Details of mixing of xanthan gum solutions
were presented in an earlier study (7).

Colloid ® solutions can be prepared by impeller stirring and
orifice shearing. Figure 3 shows that the viscosity of Colloid ®
solutions increase as the solution is passed through the orifice.
The gels formed in this case dissolve more readily than the ones
formed from xanthan gum. Consequently, even a relatively gentle
propeller stirrer can be utilized to obtain solutions which are
filterable through a 1.2 µM filter. Colloid ® solutions seem to
have good mechanical stability as will be discussed. This provides a choice of orifices with different extents of shear in the
mixer.

The difference in the dispersions of Colloid ® achieved in
propeller and orifice mixers can be seen by examining Figure 4.
In this figure the filtration times of the two samples are plotted
against the volume throughput. The orifice samples require shorter filtration times indicating better dispersion. The propeller
samples not only exhibit longer filtration times but increasingly
so. A plugging tendency is observed for these batches.

Hydroxy ethyl cellulose requires long hydration times prior
to dispersion. The formation of the gel is the critical step in

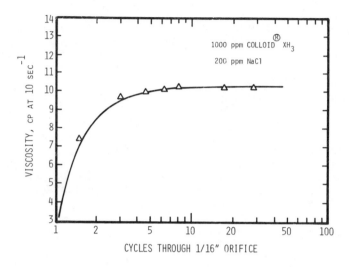

Figure 3. *Cycles through 1/16-in. orifice*

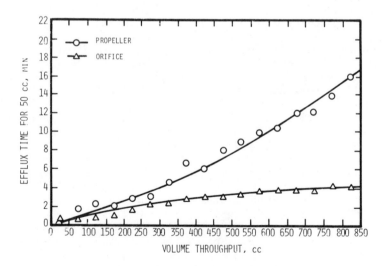

Figure 4. *Defiltration of colloid*

mixing this polymer. Once the solvent molecules are associated
with the polymer chain, the dispersion proceeds with ease (see
Figure 5). The results of propeller mixing are shown in this figure. A similar situation exists for solutions prepared by the
orifice mixer. The hydration time is necessary. Severe shearing
is not necessary for mixing hydroxy ethyl cellulose solutions,
however; at least two hours of soaking time for hydration is essential without adding a base to increase the pH. Severe shear
conditions are not detrimental to the viscosity of these solutions.

Polyacrylamides are extremely sensitive to shear fields as
pointed out in the literature (13,14). Any orifice mixing breaks
up the chains of polyacrylamides. The viscosities of Pusher ®
solutions are plotted against the number of cycles through an orifice in Figure 6. Even with a 1/16" orifice (which has less than
a 10 psi pressure drop across it), the viscosity drops from 14.1
cp to 10.9 cp showing more sensitivity than the other polymers.
Agitation with a propeller at low speeds (less than 500-600 rpm)
is a proper mixing mode for polyacrylamides.

Filtration. Many of the plugging problems associated with
flooding oil reservoirs with polymer solutions originate from in-
efficient preinjection procedures. The importance of dispersion
was presented in the preceeding section. In this discussion, the
importance and the consequences of filtration will be presented.

In general, reservoir engineers do not flood a formation with
a surfactant or polymer solution that has not been filtered. Filtration may be necessary because plugging of any portion of the
reservoir decreases oil recovered. This plugging may be caused
by polymer gels which were not completely dispersed during mixing
or impurities which act as centers for agglomerations. Proper
filtration will remove both of these undesirable species. Our
studies indicate that the most popular and effective means of filtration is diatomaceous earth. Lipton and Burnett also state this
finding (15,16). Figure 7 shows the effect of diatomaceous earth
(DE) on injectivity of xanthan gum. The degree of injectivity is
measured as the extent of decrease in flow rate observed as a certain volume of fluid is passed through a filter paper (1.2 μM in
this case) at a certain pressure drop across the paper (20 psi in
this application). The efflux time for 50 cc of unfiltered solutions exceeds 120 sec when 150 cc has passed through the filter.
The corresponding efflux time is approximately 10 seconds (even
after 1000 cc has flown through) for a solution that has been
filtered through a bed packed with 20 g of DE. The data show that
filtration through DE increases the 'injectivity' of a solution.
There are three kinds of behavior that can be observed in a DE
bed; plugging (too much throughput or extensive contamination),
ineffective filtration (too little throughput or clean solution),
and proper filtration (correct amount of filter capacity for given
throughput and right filter quality). As indicated previously,

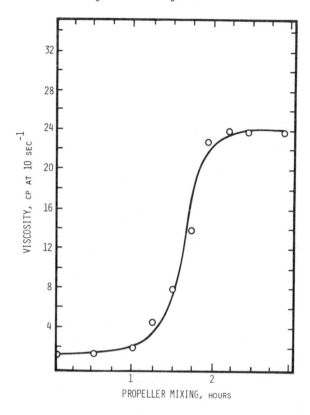

Figure 5.   Mixing of Natrosol

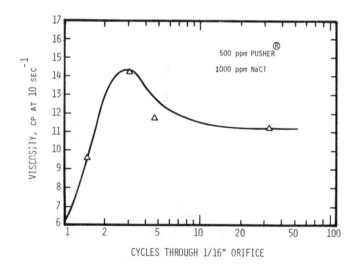

Figure 6.   Cycles through 1/16-in. orifice

solutions prepared by propeller show some extent of plugging
while the orifice samples are cleaner. The DE used was Speedflow
Ⓡ with an effective filter size of 0.50 µM in the absence of
major change in the bed permeability. The viscosities of the concentrated polymer solutions discussed in this paper refer to the
unfiltered samples.

Filterability. Effect of filtration on solution viscosities
is important because one wants to remove only a minimum number of
well-dispersed molecules while filtering out all undesired species. Filterability can be defined as the percent viscosity maintained after filtration through a paper with a small effective
pore size. Filterability of polymer solutions is shown in Figure
8. The Colloid Ⓡ samples prepared by the impeller mixer lose an
appreciable portion of their viscosities when filtered through
0.45 µM filters. On the other hand, the orifice samples maintain
the same viscosity they had prior to filtration through a 0.45 µM
filter. Filterability of xanthan gum is noted in Figure 8. Kel-
zan-XC Ⓡ does not undergo an appreciable viscosity loss when mixed
properly in an orifice blender. Hydroxy ethyl cellulose is affected by filtration quite severely. Almost all the polymer is
retained on the filter paper when a 0.45 µM effective pore size
is utilized. This may be due to the high polymer concentration
used in the solution. A higher concentration of hydroxy ethyl
cellulose is required to achieve the level of viscosity obtained
by utilizing 500 ppm of the other polymers mentioned. Polyacrylamides undergo sharp viscosity decreases even when 1.2 µM filters
are employed in the filtration. The associations between the
molecules account for this observed unfavorable filterability in
this case.
    In conclusion, these studies show that if Colloid Ⓡ and xan-
than gum are prepared in an orifice blender, the solutions can be
filtered through filters as fine as 0.45 µM without a substantial
loss of polymer. On the other hand, polyacrylamide and hydroxy
ethyl cellulose solutions lose an appreciable part of their viscosity upon a similar filtration as indicated above.

Viscosity Measurements. Low shear rates down to 1 sec^{-1} can
be observed in tight oil sands like Bradford, Pennsylvania No. 3
oil sands. To measure viscosities at these low shear conditions
accurately, a capillary viscometer has been designed and utilized
(17). The schematic of this viscometer is shown in Figure 9.
Hagen-Poiseuille law describes the flow of a Newtonian fluid
through the capillary. The Rabinowitsch correction is introduced
to obtain the shear rate at the wall for a non-Newtonian fluid
that obeys the Ostwald-de Waele model (7). A comparison of this
viscometer and the Mechanical Spectrometer Ⓡ (a sophisticated
rotational viscometer) is shown in Figure 10 (24). The good agreement between the two instruments indicates that they are both
capable of accurately determining the effect of shear rate on

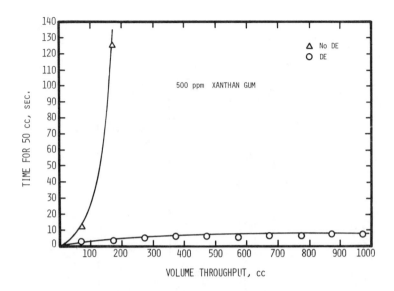

Figure 7. *Effect of DE filtration on injectivity*

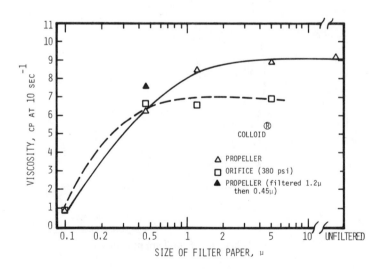

Figure 8. *Effect of filtration on viscosities*

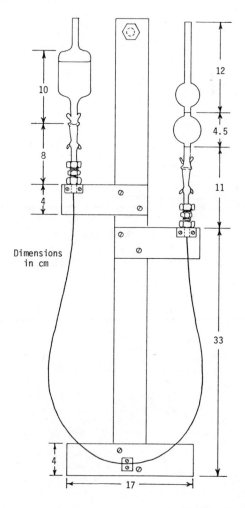

Figure 9.   Schematic of capillary viscometer (7)

viscosity. The Mechanical Spectrometer ® is an easier instrument
to operate for covering a large shear rate range but the capillary
viscometer is inexpensive. A comparison of the Brookfield ® Viscometer (which is employed in this kind of an application quite
frequently) and the Mechanical Spectrometer ® is provided in Figure 11. The main disadvantage of the Brookfield ® type viscometer
is that the viscosity over an average range of shear rate is measured, and it is not possible to determine the actual viscosity-
shear rate relationship. The Brookfield ® viscometer provides
relative viscosities while the data from the capillary unit and
the Spectrometer ® are absolute. The cone and plate geometry in
the Mechanical Spectrometer ® assures that the shear rate is
nearly constant throughout the sample. A discussion of rotational
viscometers is presented by Whitcomb, Ek, and Macosko (9).

All viscosities (except those in the temperature study) were
measured at 77°F with the capillary viscometer and they were corrected to 10 sec^{-1} if not noted otherwise. The capillary viscometer has an accuracy of 0.1% for a 10 cp fluid at 10 sec^{-1}.

Comparison of Non-Newtonian Behavior

All the polymers that have been suggested as mobility control
agents are pseudoplastic, i.e., they are shear-thinning. The extent of pseudo-plasticity is different for polyacrylamides, polysaccharides, and hydroxy ethyl cellulose. Viscosities of these
polymers are plotted against shear rate in Figure 12. One model
that describes the shear-thinning behavior observed is the Ostwald-de Waele relationship:

$$\eta = K\dot{\gamma}^{n-1} \qquad (3)$$

$\dot{\gamma}$ = shear rate, sec^{-1}

n = Ostwald-de Waele power law index

This equation predicts a straight line on a log-log plot of viscosity and shear rate with a slope of [n-1]. A horizontal line
or n = 1 indicates a Newtonian fluid. The fluids that obey the
Ostwald-de Waele model are called power-law fluids and n is referred to as the power-law index.

Polyacrylamides and xanthan gum show pseudoplasticity up to
shear rates of about 300-500 sec^{-1}. Above this shear rate, a
Newtonian zone appears. One would expect another Newtonian zone
at very low shear rates (6). This was not observed over the range
of shear rates considered in this study; however, Mungan showed
the existence of this zone for polyacrylamides (18).

For Colloid ® this low-shear Newtonian zone can be observed
since the viscosity of these solutions does not change appreciably
up to about 100 sec^{-1}. From hereon the·viscosity decreases as the
shear rate is increased. It has been noted by Mungan (18) and

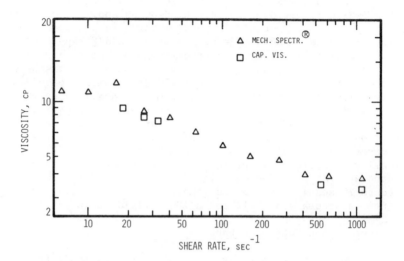

Figure 10. Comparison of Mechanical Spectrometer and capillary viscometer

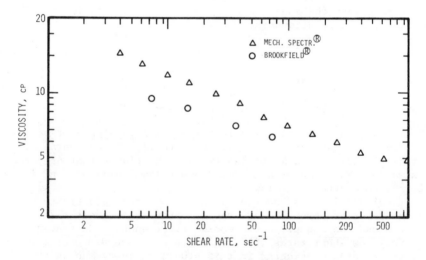

Figure 11. Comparison of rotational viscometers

Middleman (<u>19</u>) that as the molecular weight of the polymer or the average molecular weight of the solution is decreased, the low-shear Newtonian range is seen at higher shear rates. This may be why Colloid ⓇÏ which has a smaller molecular weight than xanthan gum and polyacrylamides enters the Newtonian region at higher shear rates.

Hydroxy ethyl cellulose shows a Newtonian-like behavior as shown in Figure 12. This is the case for all the concentrations of Natrosol Ⓡ investigated. The index [1-n] which indicates the deviation from Newtonian behavior is around 0.1 for NatrosolⓇ.

The xanthan gum solutions exhibit more non-Newtonian characteristics at higher polymer concentrations (see Figure 13). This is in agreement with the expectation because non-Newtonian behavior is associated with high molecular weight systems and the average molecular weight of the solution increases as the polymer concentration goes up. The correlation between [1-n] and polymer concentration for xanthan gum solutions is

$$[1-n] = 0.33 \log_{10} C_{pd} - 0.48 \tag{4}$$
$$C_{pd} = \text{polymer concentration, ppm}$$

For Colloid Ⓡ , the index [1-n] increases as the polymer concentration goes up in a similar way (see Table 3). A similar trend for polyacrylamides has been shown by Mungan (<u>18</u>).

Table 3

POWER LAW INDECES FOR DIFFERENT POLYMER SOLUTIONS

(200 ppm NaCl, ColloidⓇ-XH$_o$)

Polymer Concentration (ppm)	[1-n] (10-20 sec^{-1} range)	[1-n] (200-2000 sec^{-1} range)
125	----	0.00
250	0.13	0.13
500	0.14	0.23
1000	0.17	0.28

(500 ppm PusherⓇ -700)

NaCl Conc. (ppm)	[1-n]	Mode of Mixing	[1-n]
200	0.46	propeller	0.29
1000	0.29	90 psi - orifice	0.15
10000	0.20	380 psi - orifice	0.05

The dependency of non-Newtonian character on molecular size was mentioned above. This can be seen by examining the non-New-

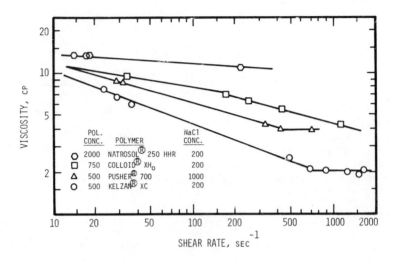

Figure 12.   Non-Newtonian behavior

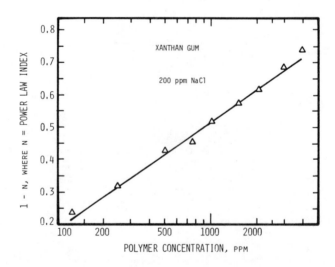

Figure 13.   Influence of polymer concentration on power law indices

tonian behavior of solutions with varying amounts of salt in them
or solutions prepared by different shears. Table 3 shows that
[1-n] goes from 0.46 to 0.20 for polyacrylamides as the salt concentration goes from 0.02% to 1.0%. The salt dependency of the
power law index of polyacrylamide solutions has been observed by
Mungan (18) also.

The effect of the mixing mode on the power law index is also
shown in Table 3. The polymer chains dissolved in a relatively
mild impeller mixer have been mechanically broken the least; consequently, these solutions have large [1-n] values. The molecules
that have been subjected to severe shearing conditions (380 psi
across the orifice) exhibit a Newtonian behavior indicating that
the polymer chains have been broken down into short segments.

From this discussion one can see that the power law indeces
obtained from Ostwald-de Waele equation for these polymers give a
good indication of the vulnerability of the chains to reversible
and irreversible changes. The ranking of the polymer solutions
according to their degree of non-Newtonian behavior is hydroxy
ethyl cellulose, Colloid ^{R}, xanthan gum, and polyacrylamides where
hydroxy ethyl cellulose is essentially Newtonian and polyacrylamides are highly non-Newtonian. This ranking is valid for fresh
solutions or solutions with low salt concentrations. Electrolytes and mechanical degradation decrease the non-Newtonian character of the solutions. Higher polymer concentrations lead to
more pseudoplasticity for all the solutions investigated except
hydroxy ethyl cellulose solutions. Power law index and polymer
concentration can be correlated as shown for xanthan gum by Equation (4).

Screen factor (a commonly used quantity) provides information
only on polyacrylamides among the four types of polymers investigated (and maybe for other polymers with high elasticity (see
Figure 14)). It is surprising that the screen viscometer data can
be correlated with oil recovery from cores because the shear rates
encountered in a screen viscometer are higher than flooding conditions by orders of magnitude.

The investigation on designing and utilizing a 'porous media
viscometer' is in progress by the authors. This viscometer will
provide information on mobility control at the low shear rate
range characteristic of flooding conditions. The properties of
the porous medium can be readily changed in this instrument.

Effect of Polymer Concentration on Viscosity

Viscosities of dilute polymer solutions increase as the polymer concentration is increased (21). Several correlations between
these two variables have been suggested. A widely used relationship is the Huggins equation which relates the viscosity to concentration in a quadratic functionality as follows:

$$\left\{\frac{\eta}{\eta_s} - 1\right\} = a_1 C_{pd} + a_2 C_{pd} \tag{5}$$

The comparison of the data and this equation is shown in Figure 15 for the four polymers investigated (the solid lines). A straight line on log-log plot of viscosity and polymer concentration also seems to fit the data well. This corresponds to a power relationship of the form:

$$\eta = b_1 C_{pd}^{b_2} \tag{6}$$

where b_2 lies between 1.1 and 1.6. In the presence of 200 ppm salt, Pusher Ⓡ solutions are the most viscous ones for a given polymer concentration; xanthan gum, Colloid Ⓡ, and hydroxy ethyl cellulose follow polyacrylamides. These comparisons are made at low shear rates.

Factors That Affect Molecular Size and Viscosity

Polymer chains in solution can undergo reversible and irreversible changes. Covalent bonds can be broken mechanically (shear degradation), by a combination of chemical and biological means (chemical degradation or bio-degradation) or thermally (high temperature depolymerization and other deteriorative reactions). Also, the chains may be taken out of solution by precipitation or adsorption, and a decrease in viscosity will result. These changes in viscosity are irreversible; the molecular weight and/or polymer concentration decreases and the solution experiences a permanent loss of viscosity.

On the other hand, there are some changes that take place reversibly. These correspond to temporary viscosity losses associated with a change in the spatial conformation of the chains in solution. Increasing the shear rate lines up the molecules in the flow field and causes the drag on the molecules to decrease. This causes a viscosity loss as discussed earlier in the paper.

The addition of electrolytes to a solution decreases the viscosity of a fresh polymer solution as will be discussed in the next section. High temperatures less than the thermodegrading conditions (less than 70-100°C depending on the polymer) cause a drop in the viscosity. This is also a reversible change associated with the change in the conformation of the chains in solution. The temperature effect on solution viscosity will be discussed. The effects of salt, temperature, severe shear, and aging will be discussed from the standpoint of viscosities and chain conformations in solution. First of all, the factors causing temporary viscosity losses will be presented.

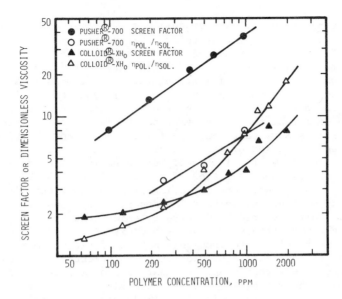

Figure 14. Screen factors

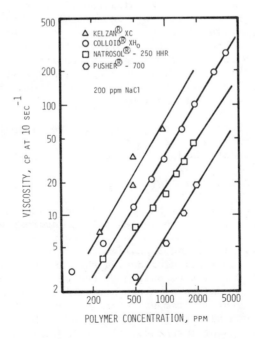

Figure 15. Viscosity vs. polymer concentration

Salt Sensitivity. The viscosity of mobility control polymers is a strong function of their environment. The ionic composition of a petroleum reservoir determines the conformation that
the polymer chains assume in it. This very fact is one of two
reasons why the salt sensitivity of the polymer solutions need to
be studied. The second incentive for an investigation of this
kind is the possibility of using the water produced from a flooded field in making the new polymer solution. Since the produced
water contains many salts and minerals, knowing how the viscosity
changes as a function of ion concentration is important.
     The viscosity of all the polymers used in enhanced oil re-
covery decreases upon addition of salt. This viscosity loss can
be related to the salt concentration as shown in Figure 16. It
is well-known that the polyacrylamides are very sensitive to salt
(20,22). Polysaccharide and polyacrylamide solutions are affected by salt more than solutions of hydroxy ethyl cellulose. This
phenomenon is related to the ionic nature of the polymer. In
fresh water, the ionic dissociation of polyelectrolytes leads to
large repulsive forces among the charged groups present in the
chain (5). These repulsive forces give rise to greatly expanded
conformations and large molecular dimensions. Electrolytes that
are in the solution shield forces and polymer size is decreased
resulting in a drop of viscosity. After all the sites on the
polymer are shielded by cations, any additional electrolytes are
relatively ineffective in decreasing the viscosity. Bivalent
cations affect the viscosity to a larger extent than monovalents
as shown on Figure 16 since they are more effective in shielding.
Polyacrylamide is the most ionic polymer among the four studied
and consequently exhibits the largest viscosity loss in the presence of electrolytes. MacWilliams, Rogers, and West showed this
by comparing the salt sensitivity of polyacrylamide solutions to
that of xanthan gum solutions (20).

     Effect of Temperature on Viscosity. The viscosity of mobil-
ity control polymers decreases with increasing temperature and an
Arrhenius type relationship is obeyed:

$$\eta = A \exp [E_a/RT] \tag{7}$$

The activation energies for the four polymers are quite different
(see Figure 17). The activation energy for hydroxy ethyl cellulose is around 5.9 kcal/gmole. For the two polysaccharides, the
activation energy is around 3 kcal/gmole. Consequently, a hydroxy ethyl cellulose soltuion having the same viscosity as
Colloid Ⓡ at 15°C will have only half the viscosity of Colloid Ⓡ
at 50°C. Polyacrylamides have two activation energies. The low
activation energy at low temperatures indicates that viscosities
do not change much with temperature. This is the case up to about
35°C. In the range of interest, polyacrylamides have the lowest
activation energy. One possible explanation might be that the

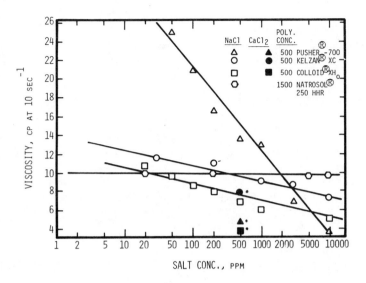

Figure 16. Effect of salt on four polymers

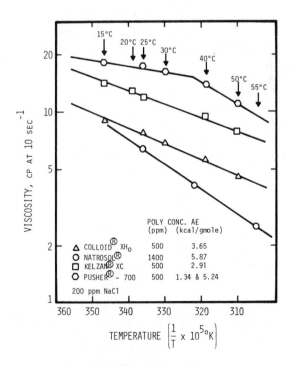

Figure 17. Viscosity vs. temperature

molecular conformation does not change appreciably at moderate
temperatures for large polymers, especially the ones with intermolecular associations.

Mechanical Degradation. All polymers have segments on the
chain that are subjected to different shear conditions in a flow
process. They undergo mechanical breakdown to different extents
depending on the kinds of bonds involved. The mechanical degradation depends to a large degree on the extent of cavitation
present (the pressure drop) and the duration and frequency of this
cavitation (passes through the orifice plate).

Mechanical degradation of xanthan gum is demonstrated in
Figure 18. Relative molecular weights are the numbers calculated
from intrinsic viscosities. At 90 psi pressure drop across the
orifice, there is little degradation. At 500 psi, severe mechanical degradation takes place and 30-40 passes through the orifices
seem sufficient for reaching a terminal viscosity level. Relative
molecular weight drops from 8 million to less than 2 million in
this case.

Mechanical degradation of hydroxy ethyl cellulose is illustrated in Figure 19. Again 100 psi pressure drop leads to some
degradation; however, 500 psi pressure drop across the orifice
decreases the molecular weight by more than a factor of 2.

The summary of the mechanical degradation data is presented
in Figure 20. Percent viscosity loss is calculated by taking the
difference of the initial and terminal viscosities and dividing
by the initial viscosity.

Polyacrylamides are broken into segments even at mild pressure drops across the orifice such as 10-50 psi. Even a propeller rotating in excess of 1000 rpm is capable of breaking up the
chainsof polyacrylamides.

Colloid Ⓡ and xanthan gum are affected by the severity of
shearing to a small extent if the pressure drop is less than 100-
150 psi. At 350-500 psi range, major degradation takes place resulting in a viscosity   loss of 30-80%. Colloid Ⓡ is the more
resistant of the two and two levels of mechanical degradation seem
to take place for this polymer.

For pressures under 300 psi, hydroxy ethyl cellulose seems to
be the most resistant polymer (see Figure 20). Since hydroxy
ethyl cellulose has the lowest intrinsic viscosity and size in
solution, this result should have been anticipated. It should be
noted that the Natrosol Ⓡ concentration used is higher than the
concentration of the other polymers. The polymer concentration
may have a major effect on the mechanical·degradation.

Long Term Stability

Polyacrylamides do not undergo biodegradation but are vulnerable to a slow chemical (oxidative) degradation. In the experiments conducted, no appreciable viscosity loss was observed over
a period of several months (see Table 4).

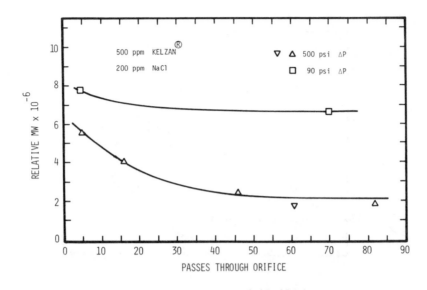

Society of Petroleum
Engineers of AIME

Figure 18. Mechanical degradation (7)

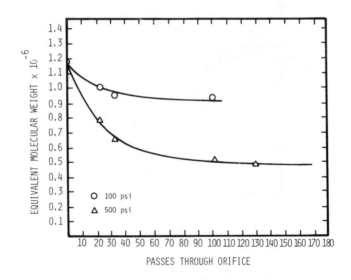

Figure 19. Mechanical degradation of Natrosol

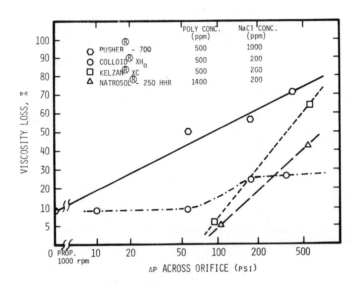

Figure 20.   Viscosity loss vs. pressure drop

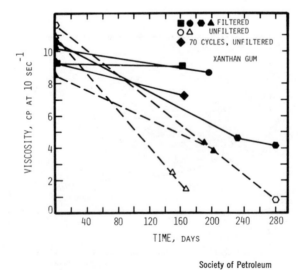

Society of Petroleum
Engineers of AIME

Figure 21.   Influence of aging on viscosities (7)

Table 4

Aging of Pusher $\circledR$

Pusher $\circledR$ Conc. (ppm)	NaCl Conc. (ppm)	Time (days)	Viscosity @ 10 sec^{-1} (cp)
1000	200	0	57.4
1000	200	155	57.4
500	1000	0	12.8
500	1000	48	11.8
500	1000	75	11.6
500*	1000	0	7.5
500*	1000	145	7.4

*Sheared in orifice.

The stability of Colloid $\circledR$ molecules is a strong function of the mode of dispersion used in dissolving them. Table 5 shows that the solutions prepared by mild agitation age slower than the ones prepared by severe shearing (380 psi across the orifice). The latter solutions contain more separate polymer segments, and hence more ends. A mechanism of aging based on adsorption or reaction of the ends would explain this observed phenomenon. Schweiger suggests that an enzymatic degradation might be responsible for this behavior (23). The conclusions concerning the filtration of xanthan gum solutions prior to aging is also valid for Colloid $\circledR$. Gel Permation Chromatography investigations of aged solutions are consistent with the viscosity measurements for Colloid $\circledR$. As in the case of xanthan gum, the injectivity profiles of Colloid $\circledR$ solutions become more unfavorable upon aging unless the solutions are refiltered.

Table 5

Long Term Stability of Colloid $\circledR$
(500 ppm CXH$_o$$\circledR$ -200 ppm NaCl)
(Unfiltered batches if not otherwise noted)

Preparation	Aging Time (days)	Visc. @ 10 sec^{-1} (cp)
Propeller	0	9.3
	47	9.3
18 psi - orifice	0	7.7
	50	7.6
55 psi - orifice	0	7.6
	30	7.4
(filtered)	52	4.8

Table 5 (continued)

Preparation	Aging Time (days)	Visc. @ 10 sec^{-1} (cp)
55 psi - orifice	0	7.7
	30	7.5
	52	3.9
180 psi - orifice	0	6.2
	51	3.3
380 psi - orifice	0	5.8
	51	2.2

Hydroxy ethyl cellulose undergoes a severe biodegradation if a biocide is not added to the solution. The biocide that performs the best is formaldehyde (see Table 6). Sodium hypochlorite is detrimental to the viscosity.

Table 6

Stability of Natrosol ® 250 HHR Solutions

Composition in Addition to 1400 ppm Natrosol 200 ppm NaCl	Viscosity (cp) Time in Weeks					
	0	1	2	3	4	5
500 ppm Sodium Benzoate	6.86	6.51	6.45	6.39	6.29	5.96
5000 ppm Sodium Benzoate	6.96	6.65	6.51	6.29	4.85	4.65
5000 ppm Mercury Chloride (HgCl)	6.82	6.50	6.30	6.18	5.90	5.45
1000 ppm Sodium Hypochlorite	5.14	1.68	1.48	1.36	0.97	0.97
5000 ppm Formaldehyde	6.86	6.66	6.67	6.75	6.82	6.31
Blank	6.87	6.22	5.27	4.01	2.66	2.60

The xanthan gum solutions that are filtered prior to aging do not lose their viscosity to an appreciable extent over a period of 6-8 months (7). Refiltering these aged solutions is necessary when determining their injectivities. A combination of biodegradation and polymer precipitation probably creates agglomerates which must be removed before injection studies. Consequently, the major change that takes place during the long term storage of xanthan gum solutions is the formation of large agglomerates which need to be filtered out but do not appreciably decrease the number of dispersed molecules in solution. Preliminary studies conducted recently on injectivities of xanthan gum solutions indicate that sodium hypochlorite improves the injectivity, but hypochlorite appears to chemically degrade the polymer molecules and hence reduce the viscosity.

Conclusions

In this paper the solution properties of a spectrum of mobility control polymers have been compared. Polysaccharides, polyacrylamides, and hydroxy ethyl cellulose show vastly different solution behavior. Despite this, the properties investigated can be correlated by noting one molecular characteristic of these polymers, namely molecular size.

Pusher ® -700 and xanthan gum have larger molecular sizes in solution than Colloid ® and hydroxy ethyl cellulose. The dimensions in solution decrease with increasing salt concentration. Polyacrylamides are affected most severely by the presence of electrolytes. Polysaccharides are also affected by salt, but not to the same extent as polyacrylamides. Hydroxy ethyl cellulose is the most insensitive polymer to salt. Temprature can be inversely correlated with viscosity. Polyacrylamides have low activation energies for viscous flow. In order of decreasing temperature dependency are xanthan gum, Colloid ® , and hydroxy ethyl cellulose.

In addition to electrolyte concentration and elevated temperatures which affect the viscosities temporarily, there are factors which alter the viscosities permanently: mechanical degradation and aging. Aging can be caused by a combination of chemical, biological, and adsorptive mechanisms.

Polyacrylamides are extremely vulnerable to severe shearing while xanthan gum, Colloid ® , and hydroxy ethyl cellulose are affected to a lesser degree.

The viscosities of xanthan gum solutions can be maintained quite well for 6 or 8 months if the solutions are mixed properly and filtered prior to aging. The degradation affects the injectability of these solutions rather than the viscosity.

Natrosol ® is sensitive to biological degradation and formaldehyde seems to be a good preservative. While sodium hypochlorite may prevent biodegradation, it has a detrimental influence on viscosities of the solution.

The viscosities of Colloid ® solutions that are prepared by gentle mixing can be maintained stable for at least two months.

Polyacrylamides are also stable for months under the conditions studied.

Dispersion and filtration are two important steps in the preparation of enhanced oil recovery agents. Xanthan gum requires severe shearing in blending. Colloid Ⓡ and hydroxy ethyl cellulose solutions can be prepared by a variety of techniques; however, hydroxy ethyl cellulose requires a significant hydration time. Polyacrylamides have to be blended gently if the molecules are not to be broken. High viscosities versus filterability is a trade-off since very viscous solutions may not be filterable through fine porous media. It may be advisable to sacrifice 10-15% of viscosity to obtain an injectable solution. Natrosol Ⓡ and Pusher Ⓡ are not as filterable as xanthan gum and Colloid Ⓡ. Pusher Ⓡ has intermolecular associations which (when coupled with high molecular dimensions) lead to this behavior. The concentration of Natrosol Ⓡ required for high viscosities are appreciable. In general, concentrated polymer solutions are not filterable.

Polyacrylamides are extremely non-Newtonian particularly at low salt conditions, while polysaccharides and hydroxy ethyl cellulose show less pseudoplasticity. The non-Newtonian behavior of all of these solutions can be described by a power law model. Many variables such as salt concentration and polymer concentration influence the extent of pseudoplasticity.

Polyacrylamides and hydroxy ethyl cellulose show widely different solution properties while polysaccharide solutions fit in between these two extremes. All these properties have been correlated.

Acknowledgements

This investigation was supported by ERDA Contract #(40-1)-5078. The cooperation of PGCOA is also acknowledged.

Abstract

Polymers that have been suggested for mobility control in oil reservoirs include polyacrylamides, hydroxy ethyl cellulose, and modified polysaccharides which are produced either by fermentation or by more conventional chemical processes. In this paper the solution properties of these polymers are presented and compared for tertiary oil recovery applications. Among the properties discussed are non-Newtonian character for different environmental conditions (electrolytes and temperature), filterability, and long term stability. The behavior of these water soluble polymers in solution can be correlated with the effective molecular size which can be measured by the intrinsic viscosity technique. A low-shear capillary viscometer with a high precision and a capability of covering low shear rates (such as 10 sec^{-1} for a 10 cp fluid) has been designed to measure the viscosities. The measurement of viscosities at such slow flow conditions is necessitated by the flooding rates in the sandstone. Mixing has been studied carefully because tight oil sands such as Bradford No. 3 require good dispersion of the polymer molecules to avoid plugging.

Literature Cited

1. Kelco - A Division of Merck & Co., Inc., "Xanthan Gum", Chicago, IL., 2.
2. Sloneker, J. H. and Jeanes, A. R., "Exocellular Bacterial Polysaccharides from Xanthomonas campestris NRRL B-1459. Part I: Constitution", Can. J. Chem., (1962), 40, 2066-2071.
3. Schweiger, R. G., Letter, Stauffer Chemical Company, San Jose, CA, October 4, 1977.
4. Hercules, Inc., "Natrosol ® - Hydroxy Ethyl Cellulose - A Nonionic Water-Soluble Polymer", Wilmington, Delaware, 1.
5. Billmeyer, F. W., Jr., "Textbook of Polymer Science", 25, 79-86, John Wiley and Sons-Interscience, New York (1962).
6. Lenk, R. S., "A Generalized Flow Theory", J. Appl. Polymer Sci., (1967) 1034.
7. Unsal, E., Duda, J. L., Klaus, E. E., and Liu, H. T., "Solution Properties of Mobility Control Polymers", Eastern Regional Conference of the Society of Petroleum Engineers of AIME, SPE Paper 6625, Pittsburgh, PA (1977), 1-5.
8. Flory, P. J., "Principles of Polymer Chemistry", Cornell University Press, London (1975), 301-309.
9. Whitcomb, P. J., Ek, B. J., and Macosko, C. W., "Rheology of Xanthan Gum Solutions", in "Extracellular Microbial Polysaccharides", Sanford, P. A. and Laskin, A., Editors, ACS Symposium Series 45, Washington, DC (1977), 160-172.
10. Lynch, E. J. and MacWilliams, D. C., "Mobility Control with Partially Hydrolyzed Polyacrylamide - A Reply to Emil Burcik", J. of Pet. Tech., (1969), 1247-1248.
11. Brandrup, J. and Immergut, E. H., Editors, "Polymer Handbook", second edition, IV-9, John Wiley and Sons, New York (1975).
12. Ibid, VI-33.
13. Maerker, J. M., "Shear Degradation of Partially Hydrolyzed Polyacrylamide Solutions", Soc. of Pet. Eng. J., (1975), 311-322.
14. Foshee, W. C., Jennings, R. R. and West, T. J., "Preparation and Testing of Partially Hydrolyzed Polyacrylamide Solutions", 51st Annual Fall Technical Conference and Exhibition of the Society of Petroleum Engineers of AIME, SPE Paper 6202, New Orleans, LA (1976), 1-11.
15. Lipton, D., "Improved Injectability of Biopolymer Solutions", Rocky Mountain Regional Meeting of the Society of Petroleum Engineers of AIME, SPE Paper 5099, Denver, CO (1975), 1-7.
16. Burnett, D. W., "Laboratory Studies of Biopolymer Injectivity Behavior Effectiveness of Enzyme Clarification", California

Regional Meeting of the Society of Petroleum Engineers of
AIME, SPE Paper 5372, Ventura, CA (1975), 1-7.

17. Liu, H. T., "A Study of Polysaccharide Solutions with a Low
    Shear Capillary Viscometer", M.S. Thesis, Department of
    Chemical Engineering, The Pennsylvania State University,
    University Park, PA (March 1977), 58.
18. Mungan, N., "Shear Viscosities of Ionic Polyacrylamide Solu-
    tions", Soc. of Pet. Eng. J., (December 1972), 470.
19. Middleman, S., "The Flow of High Polymers - Continuum and
    Molecular Rheology", John Wiley and Sons-Interscience, New
    York (1968), 150-184.
20. MacWilliams, D. C., Rogers, J. H. and West, T. J., "Water
    Soluble Polymers in Petroleum Recovery", in "Water Soluble
    Polymers", Bikales, N. M., Editor, Plenum Publication Corp-
    oration, New York (1973), 106-124.
21. Van Krevelen, D. W., "Properties of Polymers - Correlations
    with Chemical Structure", Elsevier Publishing Company, Am-
    sterdam (1972), 242, 251-59, 265.
22. Mungan, N., "Improved Water Flooding Through Mobility Con-
    trol", The Can. J. Chem. Eng., (February 1971), 33.
23. Schweiger, R. G., Personal Communication, Stauffer Chemical
    Company, October 25, 1977.
24. Wang, H. L., Personal Communication, The Pennsylvania State
    University, January 9, 1978.

RECEIVED August 22, 1978.

INDEX